THE ACOUSTICAL
FOUNDATIONS
OF MUSIC

THE ACOUSTICAL FOUNDATIONS OF MUSIC

JOHN BACKUS

PROFESSOR OF PHYSICS
UNIVERSITY OF SOUTHERN CALIFORNIA

W · W · NORTON & COMPANY · INC ·
NEW YORK

Contents

Preface

For the past ten years I have taught a course in the acoustics of music at the University of Southern California. This book brings together the material on this subject that I would like to present in such a course, but which I have never been able to because of lack of time. The students in this course are for the most part musicians with no scientific background and with no mathematical equipment past the high school algebra they have already forgotten—thus they are representative of musicians generally. It is therefore necessary to start the consideration of the acoustics of music with some elementary physics and build up a technical vocabulary as the discussion proceeds. I have tried to avoid using any technical terms (other than those in common use) without first defining them; to the extent that I have failed to do this, I submit my apologies in advance. To the extent that I have succeeded, it should be possible for the impatient reader to start on that section in which he is most interested and, as he finds unfamiliar terms, to look up their definitions in earlier chapters, and so educate himself in reverse.

Acknowledgment with thanks is due the following individuals for their permission to use published drawings and figures, as follows:

Acoustical Society of America, Fig. 7, Ch. 10;
John W. Coltman, Fig. 7, Ch. 11;
Leslie L. Doelle, Fig. 5, Ch. 9;
Harvey Fletcher, Fig. 5, Ch. 5, and Fig. 19, Ch. 11;
Carleen Hutchins, Fig. 9, Ch. 10;
Daniel W. Martin, Fig. 10, Ch. 12;
D. W. Robinson, Fig. 3, Ch. 5;
Robert W. Young, Fig. 3, Ch. 13.

In addition, I am indebted to the following for their assistance:

William T. Cardwell, for reading and commenting on Ch. 12;
John Crown, for comments on Ch. 13;
Victor Garwood, for the curves of Fig. 6, Ch. 5;
Carleen Hutchins, for the original curves of Figs. 5 and 8, Ch. 10, and for comments on Ch. 10.

Finally, I wish to express my most heartfelt appreciation to the National Science Foundation of the United States Government. The financial assistance of the Foundation has made it possible for me to pursue my own research in the field of musical acoustics. Much of what I have learned in this research has been incorporated in this book. Without the generous support of the Foundation, my work would never have progressed to the point where I felt competent to write a book on the acoustical foundations of music.

JOHN BACKUS

February 1968

Introduction

The art of music has a long and distinguished history, which parallels that of civilization itself. It is not as generally realized that the science of music has a history almost as long and distinguished. It goes back at least to the time of the Greek scholar Pythagoras (about 500 B.C.), whose name will be familiar to everyone who has taken high school geometry. From a study of vibrating strings, Pythagoras determined the conditions under which two musical tones sounded together will produce a pleasing combination. This constitutes one of the first scientific investigations in the field of musical acoustics, and will be the starting point for our later discussion of musical scales.

Since the time of Pythagoras, the science of music has engaged the attention of some of the greatest minds in the history of physics and mathematics. Aristotle, Ptolemy, Huyghens, Euler, Ohm, Young, and Helmholtz—to cite a few names familiar to physicists if not to musicians—have all contributed something toward the understanding of the acoustical foundations of music.

The scientist who becomes interested in the acoustical background of music is generally motivated by the same drive that made him a scientist in the first place: namely, an intense curiosity concerning the behavior of things, and a great delight in discovering the laws governing their behavior. He must make the assumption, rather new in human history but amply justified by this time, that the behavior of things is not decided by certain gods who (like people) are capricious, arbitrary, and unpredictable, but is governed by laws that are based on a few very fundamental principles. He assumes that these principles and laws can be worked out from observations on the way things behave under conditions that he can, to some extent, control and vary in known ways. Observations under such conditions constitute what is called *experiment*. The working out of laws and principles governing the behavior of things utilizes a complex process of experimentation, interpreting the results, constructing theories to explain the results, using the theories to suggest new experiments, and so on. This process is called the *scientific method*.

The application of the scientific method has in a very short span of human history given us an immense amount of knowledge about the things around us. This body of knowledge has been worked over and tested by a great many people, so that much of it can, with considerable confidence, be accepted as true. Much more of it can be taken as provisionally true, depending on the outcome of further scientific investigation. However, there is a great deal we do not know about the behavior of things, and the more questions we answer, the more we find remaining to be answered. Nature is infinitely complex and as our body of knowledge increases, our body of ignorance—if we may use the term—increases even faster.

The proper utilization of the scientific method to increase our body of knowledge is not easy. Significant experiments are difficult to devise; a great deal of experimentation turns out ultimately to have been useless, although undertaken for what seemed at the time to be good reasons. Even when soundly conceived, experiments are plagued by technical difficulties of various kinds; like all human efforts, they are subject to what in some circles has become known as "Murphy's Law": If anything can go wrong, it will.

Interpretation of experimental results is equally difficult. There are a great many wrong ways of explaining a given observation, and very few right ways. For example, it is the practice of some aboriginal tribes to beat on drums and other noisemakers when an eclipse of the sun occurs, in the hope of frightening off the god or demon that is swallowing the sun. Since the practice appears to be invariably successful, it might be concluded that hundreds of experiments have demonstrated that an eclipse of the sun may be brought to an end by beating on drums.

As an additional complication, it is necessary to keep in mind that what may appear at one time to be a reasonably correct explanation of certain experimental results may as a result of later research be found inadequate or erroneous; the history of science is full of examples. At one time it was certainly reasonable to suppose that the sun revolved around the earth; today this is not a very popular idea, at least among people with some education.

Because of these various problems, the scientific investigator sometimes gets the feeling that Nature enjoys placing pitfalls for the unwary, traps for the unskilled, and obstacles for everyone, to make understanding her as difficult as possible. This stubbornness develops in the investigator a tremendous skepticism. Things are seldom what they seem, as Little Buttercup sings in *H.M.S. Pinafore;* what appeared to be true yesterday may not be true today and must be tested. This skepticism must be extended to include people, especially other scientists. Anyone may be wrong about anything; no statement, made by however high an authority, is to be accepted as true simply because such a high authority has pronounced it so. Such statements may be provisionally accepted as starting points for further work, but must be abandoned without compunction if it becomes necessary.

It is in the areas where new knowledge is sought that one should most distrust Nature and prior authority. Here appearances can be most deceiving; experimental results must be checked, rechecked, and checked again, as time permits. The scientist's distrust must include even himself; he is in the unfortunate position of never being sure he is right. To guard against being led astray by preconceived notions and misconceptions, he must continually submit his work to the scrutiny of his colleagues and must at least appear to be grateful if it is torn to shreds. One of the best expressions of this point of view is given by a recent Nobel Laureate in physiology:

> [One] way of dealing with errors is to have friends who are willing to spend the time necessary to carry out a critical examination of the experimental design beforehand and the results after the experiments have been completed. An even better way is to have an enemy. An enemy is willing to devote a vast amount of time and brain power to ferreting out errors both large and small, and this without any compensation. The trouble is that really capable enemies are scarce; most of them are only ordinary. Another trouble with enemies is that they sometimes develop into friends and lose a good deal of their zeal. It was in this way that the writer lost his three best enemies.[1]

Science and its resulting technology have so tremendously changed our environment, affording us both the means for producing unlimited material goods and the means for blowing us all to eternity, that it has become imperative for the nonscientist, including the musician, to have some appreciation for the scientist's way of thinking. One way of doing this is to follow in some detail a particular area of scientific knowledge. The musician is fortunate in this respect, since his art is based in part on a foundation of acoustics, which is a portion of the field of physics; his acquaintance with science can therefore begin in an area of direct and immediate interest. In the process of learning about the science of music, the musician may acquire some of the scientist's skepticism. This will be good; an attitude which assumes that whatever anyone says is probably wrong may not be popular, but it can make things much more difficult for those demagogic types (such as certain orchestra conductors under whom the author has played) who compensate with conceit and arrogance for what they lack in ability and intelligence.

A more practical reason for the study of the acoustics of music is to acquaint the musician with the basis of his craft, and enable him to understand which physical things are important to it and which are not. He should know, for example, that a rise in air temperature affects the intonation of the winds, but not of the strings. A great deal in music is not susceptible of scientific analysis, so the musician need not be afraid of being ultimately replaced by the scientist; rather, he should welcome the attention of the scientist and hope that this attention will be productive. There is little enough known about the acoustics of music.

Furthermore, what we know about the science of music has not spread very far, particularly among musicians. As a result, when a musician tries to give an acoustical explanation for something he has observed, he is almost invariably wrong. This is not surprising since, as we have already stated, there are so many more wrong ways than right ways of explaining something. However, the consequence is that music (like other fields of human endeavor) is burdened with fallacies and superstitions it could better do without. It will be part of our task in subsequent pages to dispose of some of these fallacies. If this provokes discussion and argument from musicians who find some of their cherished beliefs under attack, this will also be good; the author can use a few good enemies.

PART I

THE PHYSICAL AND ACOUSTICAL BACKGROUND

CHAPTER ONE

THE FUNDAMENTAL PHYSICAL QUANTITIES

In order to study the science of music, it is necessary to learn some of the technical vocabulary of acoustics. This requires that we first define a number of technical acoustical terms so that we may use them properly and without ambiguity. However, acoustics is the study of systems that produce and propagate what we recognize as sound, and is based on the larger area of science called physics. We must therefore begin by learning some of the technical vocabulary of physics. The entire vocabulary of physics is quite extensive, but we will need to concern ourselves with only a small part of it; that part, however, will be quite indispensable.

The science of physics begins by considering objects and concepts with which we are intuitively familiar because we deal with them constantly in our everyday experience. However, the discipline of physics refines our thinking about these things not only by defining them as rigorously as possible, but also by making these definitions quantitative. This makes it possible to describe our objects and concepts by using numbers; they then become what we will call *physical quantities,* and we can discuss them with much more precision than things to which we cannot attach numbers.

When dealing with physical quantities as described by numbers, the use of mathematics becomes not only useful but indispensable. Mathematics is an abstract construction of the human mind, and it is really quite miraculous that it should have an immediate and practical application to the real world, serving as a quantitative language with which to discuss those things that can be described by numbers. Mathematics, like physics, covers an extensive area, but what we will need is even a smaller part than we need of physics; the barest essentials of algebra will be sufficient. Those who view mathematics as beyond understanding may be comforted by the assurance that the little we will employ will require not quite as much effort as balancing a checking account at the end of the month. What we will need is not so much the paraphernalia of mathematics as the practice it develops of thinking in quantitative terms. To try to dispense completely with mathematics and mathematical notation would be comparable to trying to study harmony without using the musical staff or musical notation.

With these preliminaries out of the way, we may begin our discussion of physics. We start by defining three fundamental physical quantities: length, time, and mass. When these are defined, other physical quantities can then be defined in terms of them. The quantities chosen as fundamental are actually selected rather arbitrarily and could be replaced by others; however, the three that have been selected serve as well as any.

Length

The first fundamental unit chosen is that of *length*. This is a physical quantity with which we are all quite familiar. It is associated with the equally familiar concept we call *distance*, the spatial separation of two points. To determine a distance, we first select some agreed unit of length and see how many times this unit is contained in the given distance. The result of this process is called a *measurement*. In general, our unit of length will fit into the given distance a certain whole number of times, with something left over. By subdividing the unit of length into smaller portions, the amount left over can be measured in terms of these subdivisions. With sufficiently fine subdivisions, the measurement of distance can theoretically be made as accurate as we please.

To be widely useful, the unit of length should be one agreed upon by the majority of users; it then becomes a *standard of length*. In the early days of human history this standard could be quite rough; the length of a man's foot, for example, could serve reasonably well as a local standard, particularly if it were a foot belonging to the local king. However, there are certain disadvantages in a standard of this kind. For one thing, each separate kingdom was likely to have its own standard; this was the case in medieval Europe.[1] For another, the accession of a new king was likely to result in the adoption of a new standard, with resultant confusion. A standard of length that is not the same everywhere and that can change from time to time makes for considerable difficulties in trade and commerce, so it was inevitable that an international standard would eventually be adopted.

Toward the end of the eighteenth century a group of scientists was engaged to devise such an international standard of length. A good standard should be one that can be independently reproduced if it should become necessary, and it was decided that one based on the dimensions of the earth would be suitable. Accordingly, it was further decided to choose as the standard of length exactly one ten-millionth or 10^{-7} of the earth's polar quadrant—that is, the distance along the earth's surface from the North Pole to the Equator. (The exponential notation used here and subsequently is explained in the Appendix.) This chosen standard was named the *meter*. A metal bar was then constructed whose length, as nearly as could be ascertained at the time, was one meter; this bar served as the world standard for about a century.

Subsequently an improved standard was constructed, the old one having become somewhat worn. The new one consisted of a bar of platinum-iridium alloy with a fine scratch near each end, the distance between these two scratches being the standard meter. It had been found that the original metal bar was not exactly 10^{-7} of the earth's quadrant, so the new standard was arbitrarily defined as one meter, without reference to the earth. This standard is kept at the International Bureau of Weights and Measures in Sèvres, France; copies of it were distributed to other countries to serve as subsidiary standards. (In 1960 this standard had become too inaccurate for present-day measurements, and the meter was redefined in other terms; however, we need not concern ourselves further with it.) The system of units based on the meter as the standard of length is called the *metric system*.

The meter serves as the standard of length all over the world. This is true even in England and the United States, which do not customarily use the meter as a unit of measurement, but instead use the *foot*. However, the foot is not based on a separate standard; it is defined in terms of the meter, the relationship being 1 foot = 0.3048 meter, exactly. The foot is the unit of length in the *English system*.

For the measurement of lengths it is convenient to have available various-sized multiples and sub-multiples of the standard. Those in the English system are quite inconveniently arranged, with 12 inches = 1 foot, 3 feet = 1 yard, 5280 feet = 1 mile, and so forth. This creates unnecessary trouble; for example, to convert a distance given in feet to the same distance expressed in miles, we must divide the number of feet by the awkward number 5280.

The metric system is much more sensibly arranged in this respect. All multiples and sub-multiples are expressed in terms of powers of 10, such as 100, 1,000, and so on. A length of 0.01 meter is called a *centimeter*, and 1,000 meters constitutes one *kilometer*. Conversions are now much simpler. To change a distance given in meters to the same distance expressed in centimeters, we multiply the number of meters by 100, and this merely means moving the decimal point; for example, 2.67 meters = 267 centimeters.

The use of prefixes to indicate the factors of 1000, 0.001, and so on, is quite useful. For example, the prefix *centi* means $\frac{1}{100}$ or 10^{-2} of whatever it is attached to, as 1 centimeter = 0.01 meter. Those we will need to know are as follows:

$$\begin{array}{lll} \text{mega} & - 10^6 & = 1,000,000 \\ \text{kilo} & - 10^3 & = 1,000 \\ \text{centi} & - 10^{-2} = & 0.01 \\ \text{milli} & - 10^{-3} = & 0.001 \\ \text{micro} & - 10^{-6} = & 0.000001 \end{array}$$

For example, one millimeter = 10^{-3} meter, and is a convenient unit for small distances. The above list is extended considerably in both directions for use elsewhere in the subject of physics, but we do not need more than the above.

To save space in diagrams and in mathematical equations, it is convenient to abbreviate the names of the units in the recommended fashion, as follows:

$$\begin{array}{ll} \text{meter} & - \quad \text{m} \\ \text{centimeter} & - \quad \text{cm} \\ \text{millimeter} & - \text{mm} \\ \text{kilometer} & - \quad \text{km} \\ \text{foot} & - \quad \text{ft} \end{array}$$

Abbreviations for names of other quantities will be given as they are defined.

The concept of length is involved in what we call a *displacement*, which we obtain when we move an object from one place to another. Physicists cannot talk for very long without drawing pictures and this is a good place to start; Fig. 1 illustrates the displacement of an object moved from point *A* to point *B*. The amount of the displacement obviously will be measured in terms of some unit of length. The physicist, who never uses a short word when a long one will do, calls the amount of a displacement its *magnitude*. A displacement is not specified completely if only its magnitude is given; it is also necessary to give its direction. We shall meet with a number of physical quantities of this sort, which involve both direction and magnitude; they are called *vector* quantities. In Fig. 1 the arrow drawn from A to B can be

FIG. 1. A displacement has both magnitude and direction.

used to represent the displacement, both in magnitude and direction, and this representation will be used with other vector quantities as needed. Most of the time we will not be concerned with the directions of displacements and other vector quantities, but only with their *sense* —that is, whether they are positive or negative.

After we have defined the fundamental units, we can combine them in various ways and obtain new ones that are called *derived* units. From the unit of length we can obtain the unit of area as a derived unit. The area of a rectangle, for example, is obtained by multiplying its length by its width; a square 5 meters on a side would have an area 5 meters × 5 meters = 25 square meters. The square meter is a derived unit. Since it is obtained by multiplying meters by meters, which is equivalent to meters squared, it is convenient to use m^2 as an abbreviation when needed. Similarly, the volume of a box, found by multiplying the area of its base by its height, would be expressed in cubic meters, abbreviated m^3. Many of the derived units we shall obtain subsequently will be used often enough to be given special names. For example, the *acre* is a unit of area in the English system.

Time

The next fundamental unit chosen is that of *time*. As with length, we have an intuitive feeling of what we mean by time; what is needed for quantitative purposes is a defined unit in which to measure it. For a considerable period of history, the rotation of the earth served as a convenient basis. We observe the sun rise in the east, cross the sky, and set in the west. We imagine a line in the sky, called the *meridian,* which runs from north to south and passes through the point directly overhead; the time required for two successive passages of the sun across this meridian is called a *solar day.* This day is divided into 24 hours (abbreviated hr), each hour into 60 minutes (min), and each minute into 60 seconds (sec); since 24 × 60 × 60 = 86,400, there are this many seconds in a day, so we may define the *second* as 1/86,400 of the solar day.

With the unit of time defined, it is possible to construct a clock and check its rate of running against the rotation of the earth by observations on the sun. The solar day varies somewhat because of the fact that the earth does not move around the sun in a circle, so observations to check the clock must be averaged over a considerable period of time. However, once the clock is running properly, it may be used to determine the number of seconds in any given time interval. Recently it has been found that the earth wobbles a bit as it rotates on its axis and is not really a good clock; as a result, it has become necessary to define the second in other terms. These problems do not concern us, and for our purposes it is sufficient to consider the second as something we can read from the second hand on a watch.

Velocity, Speed, and Acceleration

From the fundamental units of length and time, we may now obtain some important derived units that we shall need subsequently.

The first of these is again one with which we are intuitively familiar; it concerns objects in motion. For such an object, its displacement, as measured from some starting point as a reference, is continually changing. The rate at which the object's displacement is changing is called its *velocity*. Since displacement in general involves both a direction and a magnitude; so also does velocity. To describe an object in motion, therefore, we must state not only how fast it is moving, but in what direction; velocity is another vector quantity.

For our purposes, we will generally not be concerned with the directions in which things are moving, but only with the sense of the motion —up or down, for example. We may then simplify matters by considering only the magnitude of the velocity; this is a quantity we call *speed*, and one with which we are quite familiar in this automotive age.

We will measure the speed of an object by observing how many units of distance are covered in one unit of time. For example, we may observe an automobile moving along the highway and find that it travels, say, 135 meters in a time interval of 5 seconds. Its speed is then 135 meters per 5 seconds, or 27 meters per second (abbreviated m/sec). Both the amount—27, in this case—and the unit—meters per second—are equally important in describing the speed. In general, if an object moves a distance D meters in a time t seconds, its speed S in meters per second will be

$$S = \frac{D}{t}. \tag{1}$$

We could equally well measure the speed of the car in English units, in which case we would find it to be 88 feet per second. A person in the car could look at its speedometer and find its speed to be 60 miles per hour. All of these descriptions of the speed are equivalent. In fact, a sailor riding in the car would be correct in saying that its speed was 52 knots, the term *knot* being applied to a unit of speed of 1 nautical mile per hour. (Sometimes one hears the redundant term "knots per hour"; the person using it may be a sailor, but he is not a physicist.)

If the speed of an object is given, we may turn the above formula around and write

$$D = S \times t, \tag{2}$$

so that if we know the speed of an object and the time during which it moves, we can find the distance it covers. Obviously we must use consistent units; if the speed is in miles per hour, the time must be in hours, not seconds.

Both the above simple formulas are based on the assumption that the object is moving with a constant or uniform speed; if this is not so, they will not give correct answers. The corresponding mathematical formulas for the case of nonuniform speed are more complicated, and we shall not need them.

Nevertheless, the fact that the speed of an object in motion need not be constant introduces us to an important new concept. Just as velocity is the rate at which the displacement of an object is changing, so the rate at which its velocity is changing is a quantity we call *acceleration*. Since velocity involves a direction, so also does acceleration, making it another vector quantity.

The velocity of an object may change either in direction, in magnitude, or in both; in any case, there is acceleration. For example, the object may be moving in a circle with a constant speed; the direction of its velocity is then constantly changing, and it is being accelerated. However, in the instance we shall be considering, motions will generally be in straight lines, so no change in direction will be involved. Hence, for our purposes, we may consider acceleration as the rate of change of speed. If the object is speeding up, the acceleration is positive—that is, in the same direction as the object is moving. If the object is slowing down, the acceleration is negative, in the opposite sense. A negative acceleration is sometimes called a *deceleration*.

To illustrate, let us again consider the car on the road. If we push down on the pedal supplying gasoline to the motor, the car speeds up, so this pedal is sometimes called the "accelerator." If we step on the brakes, the car slows down, so we could call the brakes "decelerators." Suppose the car is speeding up, and at a particular instant the speedometer reads 15 miles per hour. A little later, its speed might be 45 miles per hour. The total change in speed is then 30 miles per hour. Let us assume that 6 seconds were required to accomplish this change. The speed of the car is then increasing at the rate of 5 miles per hour each second, or

$$\text{acceleration} = \frac{30 \text{ miles/hr}}{6 \text{ sec}}$$

$$= \frac{5 \text{ miles/hr}}{\text{sec}}.$$

This notation is rather cumbersome and since, algebraically,

$$\frac{a/b}{c} = \frac{a}{bc}, \tag{3}$$

we may write for the above,

$$\text{acceleration} = 5 \frac{\text{miles}}{\text{hr sec}}$$

$$= 5 \text{ miles/hr sec}.$$

This is still a little cumbersome, since the example we chose for purposes of demonstration has given us an expression containing two different units of time; for consistency we should use one unit or the other. Suppose we try a similar calculation with a car whose speedometer is calibrated in meters per second instead of miles per hour. Say this car's speed increases from 10 meters per second to 30 meters per second in 5 seconds; this is a total change in speed of 20 meters per second, so we get

$$\text{acceleration} = \frac{20 \text{ m/sec}}{5 \text{ sec}} = \frac{4 \text{ m/sec}}{1 \text{ sec}}$$

$$= \frac{4 \text{ m}}{\text{sec} \cdot \text{sec}},$$

meaning that the speed increases by 4 meters per second each second. However, since algebraically

$$\text{sec} \cdot \text{sec} = \text{sec}^2,$$

we may use this notation and write more concisely

$$\text{acceleration} = 4 \text{ m/sec}^2.$$

The unit of acceleration in the metric system is then 1 m/sec^2, read as "one meter per second squared."

Mass

The third fundamental physical quantity is not as easily defined as the first two, since it is not as intuitively evident. It is called *mass*. For our purposes it will be sufficient to say that mass is a property possessed by all matter. This is not much of a definition, but it will have to do. The mass of a given piece of matter is the same wherever it may be— whether on the surface of the earth or in outer space.

As with the other units, we measure masses by comparing them with a standard. The original standard of mass in the metric system was the *gram*, which was meant to be the mass of one cubic centimeter of water, taken at the temperature at which water has the greatest density. As in the case of the standard meter, inaccuracies in the early measurements made it necessary to abandon the original definition. The standard of mass is now the *kilogram* (abbreviated kg), which is defined as the mass of a particular cylinder of platinum that is kept along with the standard meter at the International Bureau of Weights and Measures. Replicas of the standard kilogram are used throughout the world as subsidiary standards.

The mass of any object can be compared to the standard mass by a process called *weighing*. We will discuss this process after we have unscrambled the concept of mass from another physical quantity that is intimately associated with it.

Force

This new quantity is again one with which we are quite familiar. It is called *force*, and is simply a push or pull. Practically everything we do in our everyday lives requires the application of forces: lifting objects, opening doors, or even simply standing. We recognize the existence of a force when it acts on our person by the physiological feeling of pushing or pulling that it produces.

In the external world, removed from any physiological sensation, a force can manifest itself in one of two ways. First, a force can produce a *distortion* in a piece of matter; that is, a force can alter the shape of an object. If we pull on a wire spring, we can stretch it out of its usual shape; if we push on a lump of putty, we can squash it into a different form. Hence, if we see a spring being stretched, or a lump of putty altering its shape, we may expect to find forces acting. The spring, if it is not stretched too far, will return to its original shape if the force is removed. Materials possessing this property are said to be *elastic*. Most substances possess this property to some degree; however, any of them will become permanently deformed if stretched too far.

Second, a force applied to a mass can produce an acceleration, setting it into motion if it is at rest or changing its motion if it is not. When we throw a ball, we certainly exert a force on it, and we observe that it accelerates until we release it. Hence if we observe a ball undergoing an acceleration, we may expect to find that a force is acting on it.

Like acceleration, a force has a direction associated with it; this was already implied somewhat in our statement that a force is a push or pull. A force exerted on a given mass may be up, down, or in any given direction; however, the acceleration produced will always be in the direction of the force. This seems fairly obvious, and can be checked in the laboratory.

There is a force which we experience constantly and which acts on every object in the vicinity of the earth; this is the force of gravity. A famous physical law, the law of gravity, states (in part) that every piece of matter in the universe exerts an attractive force on every other piece of matter. Hence the large mass of the earth exerts this force on all objects near it; the force is directed toward the geometrical center of the earth, which means that for us on the surface of the earth the force is directed vertically downward. The force of gravity acting on a mass is given a special name; it is called its *weight*. This is illustrated in Fig. 2(a); the force is represented by an arrow whose direction is

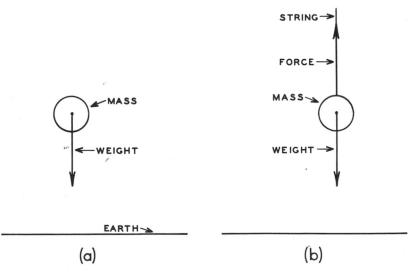

FIG. 2. (a) The weight of a mass is the force of gravity acting on the mass. (b) A mass in equilibrium.

that of the force and whose length is proportional to the magnitude of the force, as was done with other vector quantities.

Since every object on the earth has weight and since weight, being a force, can produce acceleration, we might expect that every object on the earth should be accelerated downward. This is sometimes the case, as we find when we accidentally drop something, but many objects around us are not accelerating at all. This is because the weight of the object may not be the only force acting. Suppose we tie a string to the mass of Fig. 2(a) and hold on to the string so the mass is at rest, as in Fig. 2(b). The weight of the mass is still there, and can be felt; in fact, we are pulling up on the string with a force equal to the weight. There are then two forces acting on the mass; they are equal in magnitude, but opposite in direction. Since the mass can not accelerate in two directions at once, it simply does not accelerate at all. The mass is then said to be in *equilibrium*. The two forces acting have balanced or canceled each other, so the net force acting is zero.

We can see that the equilibrium situation is a very common one. A block resting on the table, as in Fig. 3(a), is in equilibrium; the

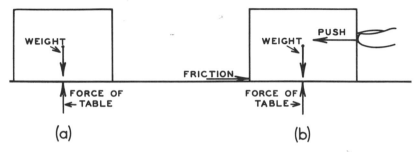

Fig. 3. (a) A block in equilibrium on a table. (b) Equilibrium with four forces.

weight of the block pulling it down is canceled by the force of the table pushing it up.

It should be mentioned that forces always occur in pairs. In Fig. 2(b), we pull up on the string to support the mass; the string pulls down on us with an equal force. In Fig. 3(a), the earth pulls down on the block, and the block pulls up on the center of the earth with an equal force. The table pushes up on the block, and the block pushes down on the table with an equal force. In any given situation, however, we are concerned with only one of the forces of each pair; in Figs. 2(b) and 3(a), we are interested only in the forces exerted on the mass and on the block, not on the forces they exert on their surroundings.

To return to the block on the table: Suppose we now push on the block in a horizontal direction. If the table were perfectly smooth, the block would accelerate along the table. However, most tables are not that smooth, and if we do not push too hard we find that the block does not move. This is because a new force appears as soon as we start to push; this force is called *friction.*

The forces due to friction are curious things. For one thing, they are always in opposition; whatever the direction the block is pushed, the frictional force pushes in the opposite direction. For another, frictional forces can adjust themselves to any magnitude within limits, so that as we push more or less hard on the block, the friction becomes whatever is needed to cancel the push. The net force on the block, both horizontally and vertically, is then zero, so the block is still in equilibrium. Fig. 3(b) illustrates this equilibrium situation, with the block at rest.

However, there is generally a limit to the amount of the force that friction can develop, so if we push hard enough, we can get the block to accelerate. If the push is adjusted to just the right value, we can get the block to slide at a constant speed, without acceleration. Here the frictional force is still equal to the pushing force, and we still have equilibrium. The criterion for equilibrium is not that the block be at rest, but that it not be accelerating. Fig. 3(b) also illustrates this equilibrium situation, the block moving with constant speed.

Forces do not have to be directed only horizontally or vertically, but can be directed at other angles; we will then have more complicated equilibrium situations. For example, suppose we push on the block on the table in a somewhat downward direction, as in Fig. 4(a), where the push is indicated by the arrow marked P. This situation is covered by a quite simple rule: we imagine the force P, with which we push the block, to be broken up into a horizontal part H and a vertical part V, as shown in the figure. These two parts form the sides of a rectangle with the original force P lying in the diagonal. The two forces H and V are then equivalent to the force P in that they produce the same effect as P; they are called its *components.* If the block is still in equilibrium, the frictional force of the table on the block must equal the horizontal component, and the vertical force of the table on the block must equal the sum of the vertical component and the weight.

Another more complicated situation is the mass of Fig. 2 hung by two strings, as shown in Fig. 4(b). The strings exert forces F_1 and F_2, which must be directed along the strings. Then the forces F_1 and F_2 are equivalent to a single force R found by another simple rule: draw a parallelogram with F_1 and F_2 as two adjacent sides; the force R is then the diagonal of the parallelogram, as shown in the figure.

So far we have said nothing about how we are to measure the force or what our unit of force is to be. Since a force can either produce an acceleration of an object or a distortion of its shape, we may in principle use either of these effects as a basis of measurement. The measurement of a force by using the distortion of an object will involve the

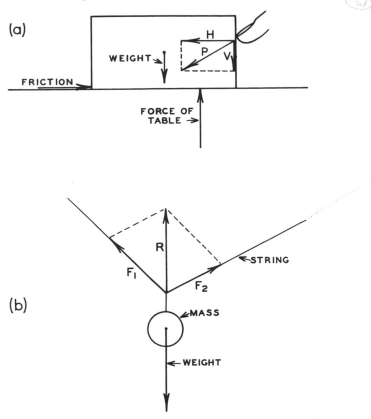

Fig. 4. (a) The block being pushed in a downward direction. (b) Mass in equilibrium supported by two strings.

elasticity of its material; this is an undesirable complication, particularly in establishing a unit which is to be used as a standard. To avoid this, forces are basically defined and measured in terms of the accelerations they produce.

To do this, we need the help of experiments in the laboratory. We assume that we already have some means of measuring forces, so that we can apply known forces to known masses and measure the resulting accelerations. If we apply various forces to a given mass, we find a quite reasonable result: the acceleration is directly proportional to the force so that doubling the force, for example, will double the acceleration. If we let a be the magnitude of the acceleration and F the magnitude of the force, we may write this

$$a \propto F. \tag{4}$$

(The symbol $\propto$ means *proportional to*.)

If we apply the same force to different masses, we find that the acceleration is inversely proportional to the mass M, so that doubling the mass will give half the acceleration. This may be written

$$a \propto \frac{1}{M}. \tag{5}$$

The mathematicians tell us that if a quantity is proportional to each of two other quantities, it is proportional to their product. Consequently, we may combine our two expressions into one:

$$a \propto \frac{F}{M}. \tag{6}$$

The next step is to change this into an equation, rather than a proportionality. We can do this by inserting a number called a constant of proportionality, so that we get

$$a = K\frac{F}{M}, \tag{7}$$

or multiplying by M,

$$KF = Ma. \tag{8}$$

The right-hand side of this equation contains the quantities mass and acceleration, which we have already defined. We have not yet defined the unit of force, and here is where we do so; to make things simple, we choose the constant K to be equal to unity and say that the unit of force is that which is defined by the equation

$$F = Ma. \tag{9}$$

If, in this equation, we make $M = 1$ kilogram and $a = 1$ meter per second squared, the force F will be 1 kilogram meter per second squared. Since we are going to use this unit a great deal, we give it a special name and call it one *newton* (abbreviated N). The newton is then that force that imparts to a mass of one kilogram an acceleration of one meter per second squared. The unit of force is now defined in terms of the three fundamental units of length, mass, and time.

A simple application of Eq. (9) above is to freely falling objects. If the string supporting the mass in Fig. 2(b) is cut, the weight of the mass will cause it to accelerate downward. This will be true for any unsupported object, for which the only force acting is its weight. If we measure the acceleration of a falling body, it turns out to be 9.8 meters per second squared regardless of the mass of the body. (This will not ordinarily be true for a light object like a feather, since air friction supplies some supporting force. In a vacuum a feather will fall as fast as any other object.) It follows from Eq. (9) that the force of gravity acting on an object is proportional to its mass, and is 9.8 newtons for each kilogram.

If a mass is accelerated by a force and the force removed when a given speed is reached, the mass will continue to move in a straight line with that speed. Only the application of another force can change the speed; in particular, the mass can be stopped only by applying a force in the direction opposite to its motion. This property of a mass in motion to remain in motion is called *momentum*.

The definition of force in terms of mass and acceleration is quite fundamental in that it does not depend on any material property other than mass. However, in practical work the use of accelerations to measure forces is rather inconvenient, so instead we may use for this purpose the distortions forces produce in material objects. For example, a coiled spring made of some elastic wire such as spring steel may be stretched by applying forces to it, and the amount of stretch is proportional to the force, provided the spring is not stretched so far as to acquire a permanent deformation. If we hang the spring from a support, as in Fig. 5(a), the bottom of the spring will be in a certain

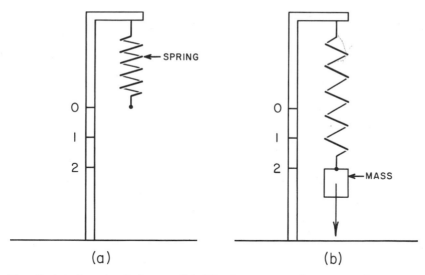

(a)   (b)

FIG. 5. (a) A spring balance. (b) Weighing a mass by means of a spring balance.

position, which we may mark as "0" on the support. A force of one newton. downward will stretch the spring to a new point which we may mark "1" on the support. A force of two newtons will then stretch it twice as far, giving point "2," and so on. This arrangement is called a *spring balance,* and is useful for practical measurements of forces.

A spring balance can be used to measure masses, since the weight of an object is proportional to its mass. If a mass is suspended from the spring as in Fig. 5(b), a position can be found for which the mass remains at rest, with the spring stretched a certain amount. This is obviously an equilibrium position, for which the weight of the mass downward is balanced by the force of the spring upward. Masses can thus be compared by the amount of stretch they produce in the spring; this is one method of weighing, which we mentioned earlier.

This method of weighing has the disadvantage of depending on the elasticity of the spring. Another method of weighing is by the use of a *beam balance,* or platform balance, as illustrated in Fig. 6. This device

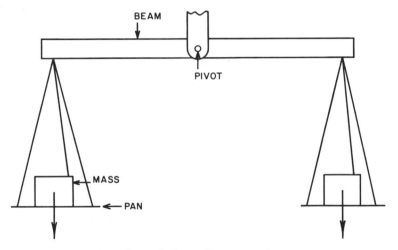

Fig. 6. A beam balance for comparing masses.

consists of two pans suspended from a beam which is supported at its center by means of a pivot. The beam will tip to one side or the other depending on which pan has the larger downward force acting on it; if the forces are equal, the beam remains horizontal. Since equal masses will have equal weights, the beam balance thus provides a method for determining when masses are equal. Masses can thus be weighed with this arrangement by comparing them directly to a standard.

On the moon, the force of gravity is about one-sixth what it is on the earth. A spring balance marked off to read weights on the surface of the earth would give quite erroneous readings if taken to the moon; its reading for a given mass would be one-sixth as much. A beam balance, on the other hand, would read correctly, since it compares masses directly.

In our discussion up to this point we have avoided the use of English units. The reason for this is that the common term *pound* (abbreviated lb) is used in our everyday lives as both a unit of mass and a unit of force, with no distinction between the two. Ordinarily this causes us no trouble, since a mass of one pound, for example, has a weight of one pound. However, if we wish to use these quantities in physical formulas, we must be careful to distinguish between pounds force and pounds mass. For example, an object falls with an acceleration of 32 feet per second squared, this being the acceleration produced by a force of one pound acting on a mass of one pound. In our Eq. (8) above, the constant K must then be 32, so the equation corresponding to Eq. (9) above for English units is

$$32\ F = Ma. \tag{10}$$

Furthermore, a man with a mass of 150 pounds and having a weight of 150 pounds on the earth would have the same mass on the moon, but his weight would go down to 25 pounds. In a satellite orbiting the earth, his mass would still be 150 pounds, but his weight would be zero. This confusion of terms makes it desirable to avoid the English system as much as possible.

Pressure

In our discussion of forces we have so far considered that they act essentially at points. The string supporting the mass of Fig. 2(b) is acting at a single point. The weight of the mass may also be considered to act at a point; if the mass is a sphere, this point is at its center. Similarly in the other examples, there is no need to think of the forces as acting at anything but points.

However, there are other situations in which a force is definitely not acting at a point, but is instead spread out over an area. In this case we can think of the amount of force that is acting on each unit area; this is a quantity called *pressure*, one we will refer to very often. The unit of pressure will be newtons per square meter (abbreviated N/m^2).

Fluids exert pressures on their containers, for example, and anything immersed in them. Fig. 7 illustrates forces due to pressure acting on the surface of a vessel containing liquid, and on the surfaces of an object immersed in the liquid. The forces due to pressure are always perpendicular to the surface on which they act.

We may illustrate the concept of pressure further with a simple example. Suppose Fig. 7 represents a square tank whose bottom measures 0.5 meters × 0.5 meters and which contains water to a depth of 2 meters. The total mass of water in the tank will then be 500 kilograms. Since each kilogram weighs 9.8 newtons, the total weight of water will

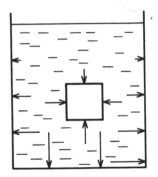

Fig. 7. Forces exerted by the pressure of a liquid on the container and on an immersed object.

be $500 \times 9.8 = 4.9 \times 10^3$ newtons. This force is spread over an area of 0.25 square meters, so the pressure p on the bottom of the tank is

$$p = \frac{4.9 \times 10^3 \ \text{N}}{0.25 \ \text{m}^2} = 1.96 \times 10^4 \ \text{N/m}^2.$$

In English units, pressure is usually expressed in pounds force per square inch. When we ask the service station attendant to put 30 pounds in our tires, we do not mean that we want a mass of 30 pounds of air in them. What we actually want is enough air to give a pressure of 30 pounds force per square inch on the inside of the tire.

We spend our lives in a sea of atmosphere, which exerts a pressure like any other fluid. This pressure amounts to very nearly 10^5 newtons per square meter, or about 15 pounds force per square inch. Normally we are unaware of this pressure, but we do notice changes in it if they are large enough, as when we go up in an airplane or under water; generally the change in pressure manifests itself in our ears as curious noises or perhaps pain. The actual value of the atmospheric pressure changes from time to time; its value at any given time is called the *ambient pressure*. Slow changes in the ambient pressure are of interest to meteorologists, furnishing information on changes in the weather. Small but rapid changes in the ambient pressure produce sensations in the ear which we call *sound*, and which we shall subsequently study in considerable detail.

Work and Energy

The terms *work* and *energy* have various shades of meaning in our everyday lives. In physics, however, they have very specific meanings. If a force acts on an object and results in the object moving, then physical *work* is done by the force. In Fig. 3(a), neither the force of gravity nor the force of the table acting on the block does work, since the block does not move. In Fig. 3(b), however, if we push hard enough on the block to move it, we will do work.

The work that is done is measured by the product of the force times the distance moved in the direction of the force. If F represents the force and D the distance moved, then the work W done by the force is

$$W = F \times D. \tag{11}$$

We could express this in newton meters but, since in physics we use this quantity a great deal, we give the unit of work a name of its own and call it a *joule* (abbreviated J). If a force of one newton acts through a distance of one meter, it does one joule of work.

The distance used in Eq. (11) above must be that moved in the direction of the force. In Fig. 3(b), for example, the force of gravity on the block does no work when the block moves along the table because there is no motion in the direction of the force.

The term *energy* means the same as work, the difference between the two words being in their grammatical usage; when we do work, we are said to expend energy. To the physicist, energy is just as real as money in the bank, because it has many of the same properties. It cannot be expended unless there is a supply to draw on. Once spent, it may sometimes be retrieved to some extent, but frequently it is gone forever.

To illustrate: If we lift a mass of one kilogram off the floor to a height of two meters, we will have to exert a force of 9.8 newtons, and so will do 19.6 joules of work. The mass can now exert forces on other objects by reason of its weight, and so can do work in turn; the amount that could be obtained is just what was given it, namely 19.6 joules. The energy that the mass has because of its elevation off the floor is called *potential energy*.

If we take the same mass and throw it, we cause it to accelerate by exerting a force on it, and so do work. If we exert a force of 9.8 newtons to throw it, and apply this force over a distance of two meters, we will again do 19.6 joules of work. Because of its momentum, the mass can exert forces on other objects, as by striking them, and so can do work on them in turn. As before, the amount that could be obtained is just what was given it, again 19.6 joules. The energy the mass has because of its motion is called *kinetic energy*.

Obviously, these two kinds of energy are interchangeable. If we raise the mass off the floor and let go of it, it falls. As it descends it loses potential energy but, since it speeds up, it gains kinetic energy. When it strikes the floor it has an amount of kinetic energy just equal to the amount of potential energy it had when released.

A third way that work can be done is in overcoming friction. When we push the block in Fig. 3(b) across the table, we do work against the friction of the block sliding on the table. Work expended in this manner is used up in producing heat in the sliding surfaces, and is irretrievably gone. It is the fate of all energy to eventually disappear as heat.

Power

When we raise the mass off the floor in the illustration above, we may do so quickly or slowly. The work of 19.6 joules done will be the same in either case. If we did it in two seconds, we would do work at the rate of 9.8 joules per second. If we required 20 seconds, we would do work at the rate of 0.98 joules per second. The rate at which work is done is called *power*. It could be measured in joules per second; however, we use the unit of power so much that we give it a special name, the *watt* (abbreviated W). An expenditure of energy at the rate of one joule per second is a power expenditure of one watt.

This unit is somewhat familiar to us since our electrical equipment is rated according to the number of watts it uses when operating. A 100-watt light bulb uses up 100 joules each second it is burning. Ten such lights would use 1000 watts, which is one *kilowatt*. To burn these ten lights for 24 hours would require $1000 \times 24 \times 60 \times 60 = 8.64 \times 10^7$ joules. The local electric company is in the business of selling electrical energy and bills us at the end of the month according to how much energy we have used. The company does not use the joule as a unit of energy, but instead uses what is called a kilowatt-hour, the amount of work done by one kilowatt expended for one hour; this amounts to $1000 \times 3600 = 3.6 \times 10^6$ joules. The ten lights above would then use 24 kilowatt-hours in the 24 hours of burning. At a few cents per kilowatt-hour, the joules the electric company sells us are quite inexpensive.

CHAPTER TWO

SIMPLE VIBRATING SYSTEMS

With the physical concepts developed in the previous chapter, we may begin to study systems of importance to the field of acoustics. We will first consider simple systems whose behavior is easy to visualize and which will help to define many useful acoustical terms. It is best if such systems can be observed directly, either in the laboratory or as demonstrations provided for the purpose. However, we will have to get along as best as we can with diagrams.

Simple Harmonic Motion

We start by considering a mass resting on a horizontal table, which for purposes of illustration we shall assume is perfectly smooth, so the mass can slide without friction. The mass is fastened to one end of a coiled spring whose other end is fastened to the table. The spring is of the kind that can either be stretched or compressed. The resulting arrangement is shown in Fig. 1(a). There will be an equilibrium position for which the spring is neither stretched nor compressed; it is marked as E under the center of the mass. Now let us displace the mass to one side, say to the right, and hold it in a new position A, as shown in Fig. 1(b). We still have equilibrium, since the force F_I with which we pull is balanced by the force F_S exerted by the stretched spring.

Now let the mass be released. Immediately, the force F_I disappears, and the force F_S is now unbalanced. Since it is due to stretching of the spring, it is directed to the left, toward the original equilibrium posi-

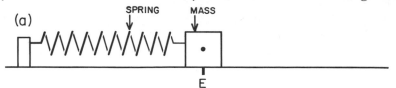

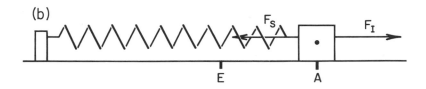

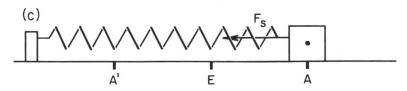

FIG. 1. A mass attached to a spring and resting on a smooth table.
(a) Equilibrium position. (b) Mass displaced and held in new position. (c) Mass released.

tion. If we had pushed the mass to the left, the force F_S would have
been directed to the right, again toward the equilibrium position. A
force of this kind that appears when an object is displaced from its
equilibrium position and is always directed toward that position is
called a *restoring force*. Furthermore, since the amount of stretch of
the spring is proportional to the force, the restoring force on the mass
in Fig. 1(c) is proportional to the displacement of the mass from the
equilibrium position.

When the mass is released at the position A in Fig. 1(c), the unbalanced force F causes the mass to accelerate and start moving toward
the equilibrium position E. At the instant it passes this position it is
moving with a certain speed, and keeps moving because of its momentum. At that instant the restoring force is zero. As soon as the mass
passes the position E the restoring force appears again, now directed
the other way. It therefore slows down the mass so that it comes to
rest for an instant at point A'. Since the restoring force is still acting,
the mass now accelerates again toward the point E. When it passes E
the second time, the restoring force brings it to rest again for an instant
at A. The whole process now starts over.

The result is a motion of the mass back and forth about the equilibrium position. This motion is of a particularly fundamental character and is basic to all the vibrating systems we will consider. The distance from the equilibrium position E to either point of maximum displacement A or A' (the two distances are the same) is called the *amplitude* of the vibration. One *cycle* of the vibration is one complete excursion from a given point, through both extremities, and back to the given point and moving in the original direction, as from E to A, back to E and to A', and back again to E in Fig. 1(c).

Since the motion repeats indefinitely, it is called *periodic*. The *period* of the motion is the time in seconds required for one cycle. For the systems we shall be dealing with, it is more convenient to use *frequency*, which is the number of cycles completed in one second, or cycles per second (abbreviated cps). We shall let f denote the frequency of a system; if T is the period, then obviously

$$f = \frac{1}{T}. \tag{1}$$

It was mentioned above that for the system we are considering, the restoring force provided by the spring is proportional to the displacement. This may be expressed as

$$F = kx, \tag{2}$$

where F is the force, x is the displacement, and k is the so-called force constant of the spring. It is a measure of its stiffness, being the force that would be necessary to stretch it one meter. When the restoring force is proportional to the displacement, as it will be for the systems we will study, it turns out that the frequency of vibration is independent of its amplitude. This is musically a most important result, since it means that whether the amplitude be small or large, the frequency of the vibration (and hence, as we will see, the pitch) will be the same.

The type of oscillatory motion we obtain when the restoring force is proportional to the displacement is given a special name; it is called *simple harmonic motion*. The reason for using the word *simple* is that any other kind of vibrating motion is more complicated to describe, and will usually be expressed in terms of a number of simple harmonic motions.

Graphs

We shall find it very useful in our discussion of acoustics to make diagrams of how some particular physical quantity depends on some other quantity, such as time. Such a diagram is called a *graph*. It is customary to plot values of the quantity we are concerned with in the vertical direction, with positive values upward from a chosen zero position. The other quantity, usually the time, is plotted horizontally, positive values to the right.

We may apply this idea to the mass on the spring and draw a graph of the displacement of the mass as it changes with time; for brevity, we may speak of the graph of displacement versus or against time. We assume positive displacements as being to the right, measured from the equilibrium position in Fig. 1(c), and conversely. We could choose the starting time as being the instant the mass is released; however, let us assume that the system has already been started, and choose our zero of time as the instant the mass goes through the equilibrium position moving in the positive direction. Assume further that the period of oscillation is four seconds; the actual value is immaterial. Then at time $t = 0$, the displacement will be zero. One-quarter cycle later, or $t = 1$ second, the displacement will be the maximum displacement A. At one-half cycle, or $t = 2$ seconds, the displacement will again be zero. At $t = 3$ seconds, the displacement will be $-A$, and so forth. If we draw horizontal and vertical lines called *axes* from which to measure distances corresponding to time and displacement, we may plot these points as shown by the black dots in Fig. 2(a). Obviously our graph must go through these points.

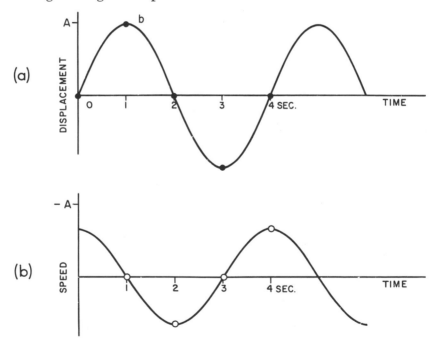

FIG. 2. (a) Graph of displacement versus time for an object moving with simple harmonic motion. (b) Graph of speed versus time.

To draw our graph accurately will require more points, and these will require measurements of the displacements at other times. We cannot do that here; however, if such measurements are made and plotted, the result will be as shown by the curve of Fig. 2(a). It is a smooth symmetrical curve; only one cycle is shown, but we may imagine it continuing indefinitely in both directions. The amplitude of the motion is shown on the graph as the distance from the highest point on the curve, such as the one marked *b*, from the horizontal time axis, and is marked by the point *A* on the vertical displacement axis. The period is given by the distance along the horizontal axis between two corresponding points. What is called the *phase* of the motion is the fraction of a cycle that has been covered, as reckoned from some chosen starting point. For example, if *0* is our starting point in Fig. 2(a), the point *b* has a phase of one-quarter cycle.

We may also make a graph of the speed of the mass versus time. We assume the speed to be positive when the mass is moving to the right. We may use the same horizontal time scale, and mark off speeds on the vertical scale. Now when the mass is at its maximum displacement, its speed is zero, since it is at rest for an instant; conversely, when its displacement is zero, its speed is greatest, being positive or negative depending on whether it is going up or down. The resulting graph of speed versus time is shown in Fig. 2(b). The two curves of Fig. 2 have the same general shape, one being displaced along the horizontal axis relative to the other.

These curves are so important that we give them a special name; we use the mathematician's description and call them *sinusoidal* curves. Similarly, any system that is vibrating with simple harmonic motion may be said to undergo sinusoidal motion.

Effect of Mass and Stiffness

In the mass-spring system of Fig. 1, let us replace the mass by a larger one, but attached to the same spring. The restoring force for a given displacement will be the same but, since it is acting on a larger mass, the accelerations will be smaller. This means that a longer time will be required for one cycle of the motion, so the period will be lengthened, which in turn gives a lower frequency. This is a general result for oscillating systems; increasing the mass of the system lowers its frequency. Mathematically, it can be shown that the frequency is inversely proportional to the square root of the mass, so that increasing the mass four times will halve the frequency.

Essentially the same argument applies if we keep the mass the same but use a stiffer spring. The system will then have a higher frequency. Again this is a general result for oscillating systems: increasing the stiffness of the system raises its frequency. If the force constant of the spring is increased four times, the frequency will be doubled.

Energy Considerations

When the spring in the system is compressed or stretched by displacing the mass in Fig. 1, work must be done, and this work is stored in the spring as potential energy. It is quite easy to show that if a spring of force constant k newtons per meter is displaced a distance x meters, the energy W joules stored is given by

$$W = \frac{1}{2} kx^2. \tag{3}$$

Hence when the mass is moved to point A in Fig. 1(c), a certain amount of energy has been given to the system. When the mass is released, it starts to move and so acquires kinetic energy. However, as it moves toward the equilibrium position, the potential energy decreases. When the mass passes the equilibrium position, all the energy is kinetic. Then, as it slows down and stops at the point A', the energy changes back to potential energy. There is thus a shuttling of energy from potential to kinetic and back again, twice each cycle. At any instant, the total amount of potential plus kinetic energy is constant. Since there is no place for this energy to go, the vibration must continue indefinitely without change in amplitude.

In our discussion of the motion of the mass on the spring, it was assumed at the start that there was no friction between the mass and the table on which it rested. This is an idealized situation; in any actual system there will be some friction, although the amount may be small. When it is present, the frictional force will use up a certain amount of energy for each cycle of vibration. As a result the energy of the system is gradually dissipated, transformed into heat. The amplitude of the vibration then diminishes and approaches zero; it is then said to *decay*. More energy is lost each cycle when the amplitude is large than when it is small; as a result, the energy disappears rapidly at first and more slowly later. The amplitude hence also diminishes in the same way. A graph of the displacement versus time for the system with some (but not too much) friction present is shown in Fig. 3. Theoretically the amplitude never gets exactly to zero, but practically it gets close enough to be considered zero. The dotted curve in Fig. 3 connecting the successive peaks is called the *envelope* or *decay curve* of the vibration.

An oscillation whose amplitude is diminishing in this way is said to be *damped*. That part of the system that absorbs energy—the friction in the system we have been discussing, for example—is frequently termed the *resistance*, or sometimes *damping resistance*.

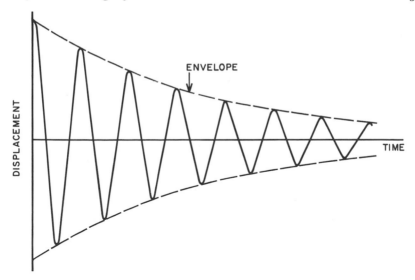

DISPLACEMENT

ENVELOPE

TIME

FIG. 3. Graph of displacement versus time for a damped vibration.

Suppose now we set this system with friction into oscillation as before; then, after it is going, we feed a little bit of energy into it each cycle. We may do this by giving the mass a little push each time it passes the equilibrium position, thus adding a little kinetic energy, or alternatively we may pull it out a little farther each time it comes to rest at the ends of the path, thus adding a little potential energy. If each cycle we add just the amount of energy that is lost to friction, the amplitude will remain constant. Since there are a certain number of cycles each second, and a certain energy given each cycle, we have a certain total energy furnished each second, so power is being supplied to the system. In this case it is supplied just as fast as it is used up, so the amplitude does not change.

Now we may do one last experiment with our system. Instead of giving it a lot of energy all at once by displacing it a considerable distance, let us start it from rest at the equilibrium position by giving it a very small push so it vibrates with small amplitude, and then start supplying power to it, a little each cycle, as above. At first the energy supplied each cycle will be larger than the energy lost to friction, so the amplitude will increase. As it gets larger, the friction losses also increase and, when the power lost to friction equals that supplied, the amplitude remains constant. This behavior is illustrated in Fig. 4, in which power is assumed to be supplied to the system at time $t = 0$.

At time $t = T_1$ in Fig. 4, the power is assumed to be cut off. The vibration then dies away as already described above. Fig. 4 thus gives a picture of the growth and decay of a vibration. The growth—that

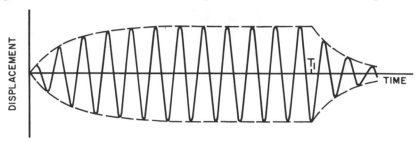

FIG. 4. Growth and decay of a vibration.

part of the curve where the amplitude is increasing—is sometimes called the *initial transient*. That part where the amplitude is constant is called the *steady state*. We shall see that the envelope of the vibration amplitude, as shown by the dashed curves of Fig. 4, is applicable to the tones produced by many musical instruments.

The Tuning Fork

We have spent some time discussing physical systems which have no apparent relation to anything musical. This has been done for the purpose of establishing a vocabulary and illustrating physical concepts. Now we are equipped to discuss a simple system that is definitely of interest to music; this is the *tuning fork*. This device consists of a metal bar bent in the shape of a letter U with a handle fixed to the curved part, as illustrated in Fig. 5(a). The prongs have mass, and the curved

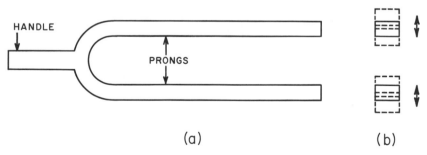

(a)   (b)

FIG. 5. (a) Tuning fork. (b) End view of vibrating fork.

portion has a stiffness which supplies a restoring force if the prongs are displaced, so the system can vibrate. It is made to do so by displacing the prongs and suddenly releasing them; one way of doing this is to strike the side of the fork with a reasonably soft hammer. The prongs vibrate in the plane of the fork, as shown in Fig. 5(b), always in opposite directions.

When the fork is vibrating, the motion of the prongs is sinusoidal. The simple experiment shown in Fig. 6 demonstrates this. One prong

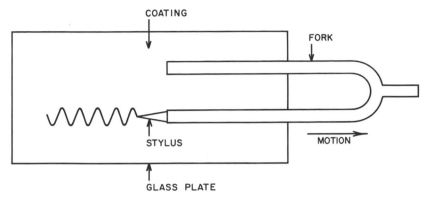

FIG. 6. A tuning fork drawing a graph of its own vibration.

of the fork is provided with a light pointed stylus as shown. A glass plate is coated with a layer of soot or other material which will yield a fine line when the tip of the stylus is drawn across it. The fork is set into vibration and the vibrating stylus drawn across the plate by moving the fork in the direction of the arrow. The stylus then inscribes a line in the coating which is found to have the shape of the graph of Fig. 2(a). Strictly speaking, the graph should show an amplitude decay like that of Fig. 3 above, since the vibrational energy given the fork by striking it or otherwise is gradually dissipated because of frictional forces. However, for a well-made fork the decay rate is quite low, and the change in amplitude is imperceptible over only a few cycles.

The tuning fork is a very convenient frequency standard. Once adjusted, a good fork will maintain its frequency quite exactly for an indefinite length of time. The frequency of the fork may be raised by removing material near the ends of the prongs; this decreases the mass of the prongs and increases the frequency, as discussed above. Conversely, removing material near the base of the prongs decreases the stiffness of the fork and lowers the frequency.

The tuning fork is commonly considered to have only one vibration frequency. Actually it has some other (undesirable) frequencies, which we shall discuss later.

CHAPTER THREE

WAVES AND
WAVE PROPAGATION

To develop a topic in physics, it is frequently necessary to adopt a somewhat circuitous course. This is the case in our discussion of acoustics, which we defined as the study of the production and transmission of sound. As we will see, sound is produced by vibrating systems, and we must describe the properties of such systems before we can discuss how their vibrations may be transmitted from one place to another through some intervening material substance. However, vibrating systems are also made up of material substances, so to discuss them thoroughly requires that we must already know how vibrations are transmitted through materials. Our next step then, is, to see how vibrations from a source such as a tuning fork can be transmitted from one place to another; we can then come back and discuss vibrating systems more completely.

The Medium

To transmit vibrations from one place to another requires that there be some material in the intervening space. This material is called the *medium*. It may be matter in any form—solid, liquid, or gas. It may be essentially one-dimensional, as for example a stretched string. It may be two-dimensional, as a stretched membrane on the head of a drum or the surface of a body of water. It may be three-dimensional, as the interior of a body of water or, in particular, the atmosphere around us. The string, the membrane, and the atmosphere are all important to music.

Whatever their form, all the media with which we are concerned have certain properties in common. Since the medium is generally of indefinite extent, we do not know its total mass, nor are we interested in it: The important quantity is the mass of a unit amount of the medium; this is called its *density*. For the one-dimensional medium such as the stretched string, the density will be measured in kilograms per meter. For the two-dimensional system, it will be kilograms per square meter, and for three dimensions, kilograms per cubic meter.

Second, all the media with which we will be concerned will be elastic. All parts of the medium will normally be at rest, in equilibrium. However, if any part of the medium is displaced from its normal position by some external agency, a restoring force immediately appears that tries to push it back again.

Now suppose we create a sudden disturbance at some point in the medium, such as by quickly plucking the string or striking it with a hammer, striking the drumhead with a stick, or clapping our hands. A sudden disturbance of this kind lasting only a short time is called an *impulse*. When a disturbance is created, the displaced parts of the medium exert forces on the adjacent parts, causing them to move away from their equilibrium positions. They in turn cause the next portions to move, and so on. The disturbance then travels through the medium; this process is called *propagation*. Because of the mass of the medium, its parts do not instantly move away from their equilibrium positions when pushed by the neighboring parts, but instead require a little time to do so. As a result, the disturbance travels through the medium with a definite speed. The speed of propagation depends on the density and elastic properties of the medium.

A simple one-dimensional medium whose properties can easily be visualized is shown in Fig. 1; it consists of a number of identical masses

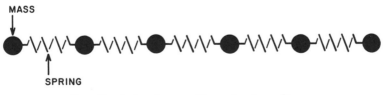

MASS

SPRING

Fig. 1. A simple one-dimensional medium.

connected together by light springs. The whole system may be imagined to extend indefinitely to the right. (We may assume the masses are supported in some frictionless manner so that their weight is not a complicating factor.) Now assume the first mass is suddenly dis-

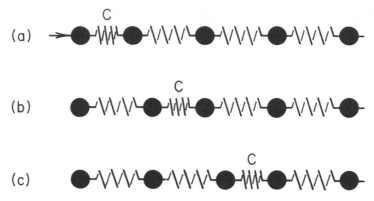

FIG. 2. (a), (b), (c). Successive positions of a longitudinal disturbance traveling in the one-dimensional medium.

placed to the right, as in Fig. 2(a). This compresses the first spring, so that it exerts a force on the second mass. This mass will then move to the right, in turn, as in Fig. 2(b), compressing the second spring, which then exerts a force on the third mass, and so on. The disturbance then travels along the medium, with successive configurations of the system shown in Fig. 2(a), (b), (c). Since the displacements of the successive masses are in the direction the disturbance is traveling, this is called a *longitudinal* disturbance.

The speed of propagation will evidently depend on the medium. Larger masses would move more slowly under the forces produced by compressing the springs, so the propagation speed would be reduced by increasing the masses. Conversely, stiffer springs would exert greater forces and cause the masses to move more quickly, and so increase the speed.

A different kind of disturbance could be created in the system. Suppose we displace the first mass upward, as in Fig. 3(a). The spring will then exert a sidewise force on the second mass, pulling it to one side in turn. It will then pull the third mass, and so on. This disturbance will travel along the system as shown in Fig. 3(a), (b), (c). The displacements of the masses are now perpendicular to the direction the disturbance is traveling; this is called a *transverse* disturbance.

Such transverse disturbances can only exist in a medium in which one part can exert a sidewise force on a neighboring part. This is possible only in solids. In liquids and gases only forces like the direct pushes and pulls of the springs of Fig. 2 can be exerted by one part on another. Disturbances traveling in liquids and gases are thus always longitudinal; this will include sounds in air.

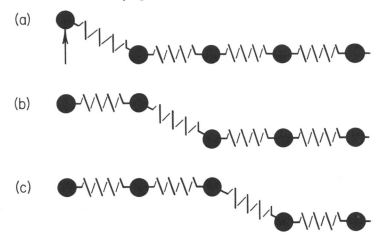

FIG. 3. (a), (b), (c). Successive positions of a transverse
disturbance traveling in the one-dimensional medium.

Waves

We shall generally be concerned with vibrating sources, such as the
tuning fork, that produce periodic disturbances following one another
through the medium, generating what is called a *wave*.

To visualize waves, let us imagine our one-dimensional mass-and-
spring medium attached to one prong of a tuning fork, as shown in
Fig. 4, so that the direction of vibration of the prongs will be perpen-

FIG. 4. Tuning fork attached to the one-dimensional medium to generate
transverse waves.

dicular to the direction of the medium. Now if we set the fork into
vibration, we will get a series of disturbances traveling down the
medium; if we wait long enough for the first one to get out of sight
on the right, we may draw diagrams showing the successive configurations of the system at various phases of the vibration of the fork.
Four of these diagrams are shown in Fig. 5 for four successive positions
of the fork prong. In Fig. 5(a) the prong is at the equilibrium position,
moving down. In Fig. 5(b) the prong is at its greatest displacement
downward and reversing direction; in Fig. 5(c) at the equilibrium
position moving up, and in Fig. 5(d) at its greatest displacement upward. The fork is then producing a *transverse wave* in the medium,

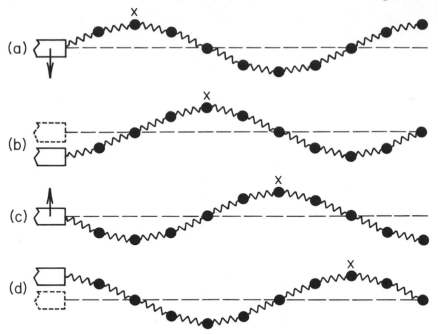

FIG. 5. A transverse wave moving in the one-dimensional medium.

consisting of a succession of *crests* and *troughs* traveling to the right, as shown by consecutive positions of the crest marked X in Fig. 5. The succession of crests and troughs in the medium is sometimes called a *wave train.*

If we turn the fork around so the direction of the motion of the prong is in the direction of the medium, as in Fig. 6, we will produce a *longitudinal wave,* consisting of a succession of alternate compressions and expansions traveling in the medium. These are shown for succes-

FIG. 6. Tuning fork attached to the one-dimensional medium to generate longitudinal waves.

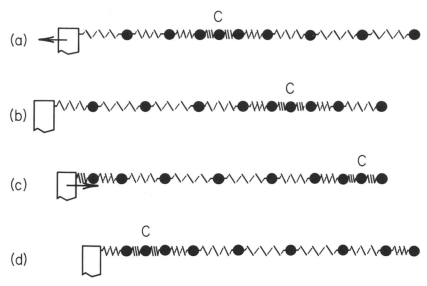

Fɪɢ. 7. A longitudinal wave moving in the one-dimensional medium.

sive phases of the fork in Fig. 7(a) through (d), as was done for the transverse wave. The wave is traveling to the right, as shown by the successive positions of the compression marked *C* in Fig. 7.

The medium we have used to illustrate these waves is rather artificial, so let us look at some actual ones. One of great practical importance to music is the stretched *string*, which is a long piece of material of essentially negligible thickness, completely flexible, and pulled apart at its two ends by equal and opposite forces which keep it taut; the magnitude of the pulling force is called the *tension*.

If we fasten a long string to our tuning fork, keep it stretched, and start the fork vibrating, we will produce transverse waves which travel along the string. If we draw diagrams of these waves at successive instants, as before, we will get those shown in Fig. 8(a) through (d). These are simply smoothed-out versions of the curves of Fig. 5. The wave is sinusoidal in shape, which is not surprising, since it is produced by the sinusoidal vibration of the fork. The dotted line in each of Fig. 8(a) through (d) represents the equilibrium position of the string when the fork is not vibrating.

Another one-dimensional system of great musical importance is the *air column*. This is simply the air contained in a tube. (The cross section of the tube is usually circular, although this is not essential, and some other gas besides air could be used.) The material of the tube

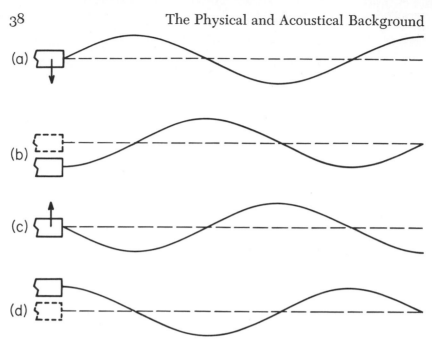

FIG. 8. A transverse wave in a string.

does not matter as long as it is rigid enough not to yield appreciably if the pressure inside it increases.

Air is a particularly important medium. Being matter, it has mass. A given mass of air occupies a volume that depends on the pressure acting on it; increasing the pressure decreases the volume, pushing the air molecules closer together, and conversely. Air is thus what is called a *compressible* medium.

Suppose we now attach a very light piston to one prong of our tuning fork and use this piston to close off one end of a long tube, as shown in Fig. 9; the piston is loose enough to slide freely in the tube. The air in the tube consists of individual molecules, represented in Fig. 9 by the black dots. (Only a few are so represented; in a cubic millimeter of air there are about a million times as many molecules as there are people on the earth.) On the average, the molecules of air are uniformly distributed throughout the tube. If we now cause the tuning

PISTON

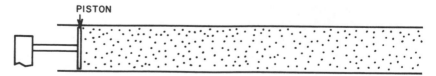

FIG. 9. Tuning fork attached to piston at the end of an air column.

fork to vibrate, the piston will alternately pull and push on the air in the tube, producing alternate *condensations* (places where the air molecules are pushed together) and *rarefactions* (where they are pulled apart), which travel down the tube as a longitudinal wave. Four successive representations of this wave at four phases of the fork vibration are shown in Fig. 10(a) through (d); the condensations and

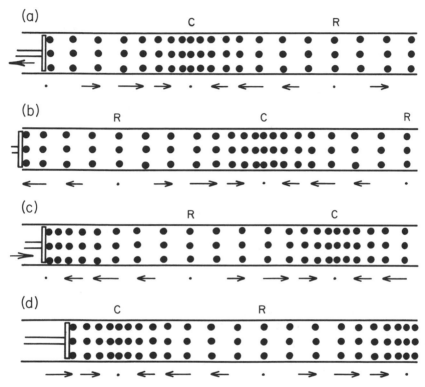

FIG. 10. A longitudinal wave in an air column.

rarefactions are marked *C* and *R* respectively. As before, only a few molecules are represented, and these may be thought of as being evenly spaced when the air in the tube is undisturbed. The displacements of the molecules when the wave is created then show up more clearly. The condensations and rarefactions are traveling to the right, as shown. The similarity of the air column waves in Fig. 10 to the waves in the mass-spring system of Fig. 7 is obvious. The longitudinal wave traveling in the air column represented by Fig. 10 constitutes a one-dimensional *sound wave*.

The longitudinal wave is harder to visualize than the transverse wave, so let us plot a graph of the displacements of the air particles versus distance along the column. Underneath the columns in Fig. 10 arrows are drawn giving at successive points along the column the displacements of the air molecules from their usual positions. If we select the instant of time represented by Fig. 10(a) and plot on a graph these displacements in the vertical direction versus distance in the horizontal direction, we get the graph shown by the solid line in Fig. 11. It is another sinusoidal curve, identical to Fig. 8(a) for the transverse wave in the string. (The horizontal line obviously represents the undisturbed medium.) The representation of Fig. 11 may thus be

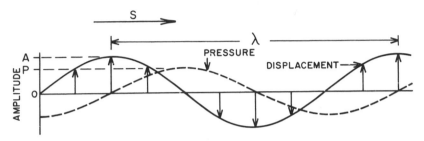

FIG. 11. Graph of the displacements of the air molecules and the pressure variation, plotted against distance along the column.

used for both transverse and longitudinal waves, and this will be very helpful in subsequent discussions.

We may also plot the pressure in the wave on the graph of Fig. 11. From Fig. 10 we see that the pressure will be higher than the average or ambient pressure at the condensation C, where the molecules are closer together. Conversely, the pressure will be lower at R. If we plot the pressure variation along the wave, we get the dotted curve shown in Fig. 11. The pressure is greatest or least where the displacement is zero, and is equal to the average pressure where the displacement is largest. The pressure and the displacement are said to be *out of phase*, in this case by one-quarter cycle. What is called the *sound pressure* is the amplitude of the pressure variation in a sound wave, above and below the ambient pressure, shown as OP in Fig. 11. (The physicist calls this the *peak* sound pressure, but we do not need to be quite so precise.)

Certain quantities relating to waves now need to be defined. First we note that the individual particles in the medium through which the wave is traveling oscillate back and forth about their equilibrium positions as the wave passes by; they do not undergo any permanent displacement. We may see this in Fig. 8 or Fig. 10. The *amplitude* of the wave is the amplitude of vibration of the individual particles, and is given by the distance *OA* in Fig. 11. It depends on the amplitude of vibration of the source of the wave, such as the tuning fork in this case. The frequency of the wave is the vibration frequency of the individual particles of the medium, and is also the number of waves per second that pass by any point in the medium. It is also determined by the source; if a tuning fork has a frequency of 100 cycles per second, it will produce 100 waves per second, and this must be the number per second that an observer stationed somewhere in the medium would see going by.

On the other hand, the speed of the wave (indicated by S in Fig. 11) depends on the medium, as we have already stated. For the string and air column, it is independent of frequency. (There are some systems for which this is not so; we will meet them later.) The speed is also independent of the amplitude of the wave. (This is true provided the amplitude is not too large. To avoid endless qualifications, let us say now that practically every statement in this book is subject to limitations beyond which it may not be true.) For the stretched string, the speed S of a wave in meters per second is given by

$$S = \sqrt{\frac{T}{\rho}}, \tag{1}$$

where T is the tension in the string in newtons and ρ is its density in kilograms per meter. We will discuss the speed of waves in air and other gases later.

Another quantity of importance in wave motion is the *wavelength*. This is the distance from any point in a wave to the corresponding point in the next wave; for example, the distance from one crest to the next, as indicated in Fig. 11 by the Greek letter λ.

The frequency and speed of the wave are independent quantities, as already shown. The wavelength, however, is not an independent quantity. If f is the frequency, then in one second there will be f waves formed by the source, each of length λ, so they will occupy a distance $f\lambda$ along the medium. This will be the distance traveled by the first wave in one second; but this is simply the speed S. For example, suppose the source has a frequency of 7 cycles per second, and so produces 7 waves per second. If each of these has a length of 2 meters, then in 1 second the front of the first wave has traveled 14 meters, as we see in Fig. 12. The speed of the wave is then 14 meters per second. Hence we find the very important relation

$$S = f\lambda \tag{2}$$

which relates speed, frequency, and wavelength, and which applies to any wave motion. This expression will find considerable application.

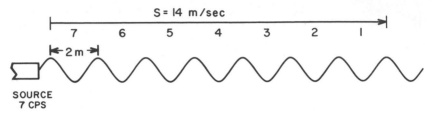

FIG. 12. Speed = frequency × wavelength.

Two-Dimensional Waves

The one-dimensional waves we have discussed at some length are of considerable practical importance, but we may also have waves in two and three dimensions. The waves that travel on the surface of a body of water furnish a very familiar example of two-dimensional waves. Water waves are rather complicated, since they are a combination of longitudinal and transverse motions, and their speed depends on their wavelength; however, they are very useful for demonstrating some other properties of waves. They are also easily visualized and can be well represented by diagrams on paper.

Suppose we put a plunger into the surface of a still pond and then move it up and down with some given frequency. The plunger will then be a source of waves which will move out in all directions along the surface of the water. We may picture these waves by imagining lines drawn through coresponding parts of successive waves, such as the crest of each wave. Such a line is called a *wave front*. Figure 13(a)

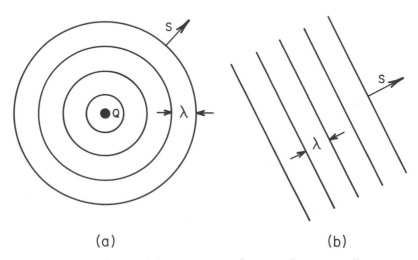

(a)   (b)

FIG. 13. Wave fronts. (a) Waves spreading out from a small source. (b) Plane waves.

shows a succession of wave fronts traveling out from the source Q. If the speed is the same in all directions, the wave fronts are circles with their centers at Q. Successive wave fronts are obviously one wavelength apart, and they are moving outward from the source, perpendicular to themselves, with the speed S.

As the wave fronts move out, their curvature becomes less, and at a great distance from the source they become essentially straight. We then have a series of *plane waves,* as shown in Fig. 13(b), where the wave fronts extend indefinitely in both directions.

Sound Waves

In an extended medium we can have three-dimensional waves. A solid will transmit both longitudinal waves and transverse waves; such waves traveling in the earth are known as earthquakes. As already stated, liquids and gases will only transmit longitudinal waves; as in the one-dimensional case, these are variations in pressure above and below the ambient pressure and so are also called *sound waves.*

The tuning fork vibrated in air will generate sound waves that are transmitted in all directions. (When we set the fork into vibration we are sometimes said to *sound* the fork; the word can serve both as a noun and as a verb.) As in the one-dimensional air column, the speed of these waves is independent of their frequency and (within limits) of their amplitude. The speed c of sound waves in a gas is given by a very simple expression

$$c = \sqrt{\frac{\gamma p}{\rho}}, \tag{3}$$

in which γ is a constant that depends on the gas and has the value 1.4 for air, p is the ambient pressure in newtons per square meter, and ρ is the density in kilograms per cubic meter.

For air at what are called *standard conditions,* namely at a temperature of 0° centigrade (32° Fahrenheit) and a pressure of 1.013×10^5 newtons per square meter (normal atmospheric pressure), the speed of sound is 331.5 meters per second (1087 feet per second).[1] Our concert halls are usually warmer than this, however, and at higher temperatures the speed of sound is greater. This is because gases expand as the temperature is increased, the pressure being kept constant, so the density decreases. The increase in speed of sound amounts to 0.6 meters per second for each degree centigrade temperature rise (1.1 feet per second for each degree Fahrenheit). For example, at a room temperature of 20°C. (68°F.) the speed would be (rounded off to three figures)

$$332 + 0.6 \times 20 = 344 \text{ m/sec,}$$
or
$$1087 + 1.1(68 - 32) = 1130 \text{ ft/sec.}$$

(This method is good only if temperature changes are not too large.) The fact that the speed of sound increases with temperature is of considerable practical importance to wind and brass players, as will appear.

The speed of sound does not depend on the ambient pressure. At first sight it would appear that it should, since the pressure appears in Eq. (3) above. However, if the pressure of a gas is increased, the density increases in proportion if the temperature is kept constant, so the ratio p/ρ in Eq. (3) stays the same, and the speed is unchanged.

Other gases have densities differing from that of air, so the speed of sound waves in them will be different. Hydrogen, for example, is a very light gas, and sound waves in it travel almost four times as fast as they do in air. In liquids and solids the speed of sound is quite large; this is because the elasticities of these materials are very much greater than that of air and more than compensate for the fact that their densities are also greater. Table I gives values of the speed of sound in various substances.

TABLE I

Speed of sound in various substances

SUBSTANCE	TEMP. °C.	SPEED m/sec	SPEED ft/sec
Air	0	331.5	1087
Air	20	344	1130
Hydrogen	0	1270	4165
Carbon dioxide	0	258	846
Water	15	1437	4714
Steel	—	5000	16,400
Helium	20	927	3040
Water vapor	35	402	1320

Doppler Effect

It was stated above that the frequency of a sound wave is the same as the frequency of the source. There is an apparent exception to this that occurs when the source producing the waves and the observer measuring the frequency of the waves are moving relative to one another. Most of us have had the experience, while riding in a train, of hearing the pitch of the bell on a roadway-crossing signal suddenly drop several semitones as we pass quickly by. In other words, the frequency of a sound appears to be higher if we are moving toward the source, and lower if we are moving away. This frequency shift is called the *Doppler effect*.

It is explained quite simply. Suppose a source has a frequency of 100 cycles per second. An observer at O in Fig. 14(a) will then count 100 waves passing per second. Now suppose the observer is moving toward the source, so that in one second he moves from O to P. He will then count not only the 100 waves that pass by O in this time but also the number in the distance O to P, so the frequency appears to be higher. Conversely, if the observer moves away from the source the frequency appears to be lower.

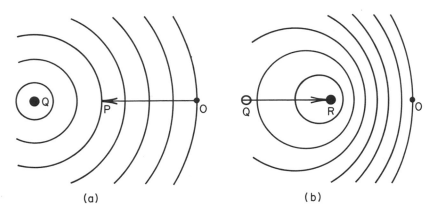

(a)   (b)

FIG. 14. Doppler effect. (a) Observer moving toward the sound source. (b) Source moving toward the observer.

The Doppler effect also appears if the sound source is moving and the observer is stationary. We have an example of this when the ice cream wagon comes along the street playing its monotonous tune to attract the children; as it passes by, the pitch drops. What happens in this case is shown in Fig. 14(b). Suppose again that the source has a frequency of 100 cycles per second. If in one second the source moves from Q to R, the 100 waves it produces are crowded into the distance RO instead of the original distance QO. The wavelengths are consequently shorter, so from Eq. (2) above, the frequency must be higher.

The amount of frequency shift produced by the Doppler effect depends on how fast the observer or the source is moving relative to the speed of sound. It is of no particular musical significance; a listener running as fast as possible toward the orchestra during the performance might just be able to hear the shift in frequency.

Properties of Waves

All wave motions have certain properties in common.

1. *Reflection.* Whenever waves traveling through a medium reach a boundary surface where the properties of the medium suddenly change,

the waves may undergo *reflection*. A rigid wall provides such a boundary surface, and sound waves will reflect from it. Figure 15(a) shows plane waves reflecting from a flat wall. The wave moving toward the wall is called the *incident* wave. In Fig. 15(a), the incident wave fronts are shown by the solid lines, moving in the direction of the arrow *I*, and the reflected wave fronts are shown dotted, moving in the direction *R*. The amount of sound reflected depends on the nature of the wall; we will discuss this more fully below.

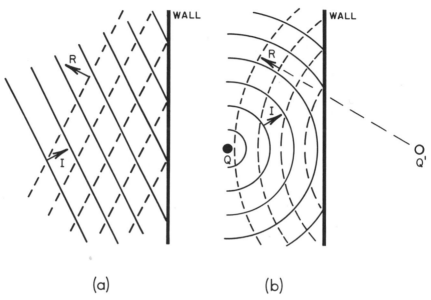

(a) (b)

Fig. 15. (a) Reflection of plane waves from a flat rigid wall. (b) Sound source in front of a wall and its image behind the wall.

A sound source *Q* placed in front of a flat wall is shown in Fig. 15(b). The incident waves spreading out from the source and moving to the right are, after reflection from the wall, moving to the left, and appear to be coming from a point *Q'* behind the wall. This point is called the *image* of the source; it is as far behind the wall as the source is in front. If we stand a distance off from a wall and shout, the sound waves reflect from the wall just as though the image had shouted at the same time; they then return to us delayed by the time interval it would take them to come from the image. This produces what is called an *echo*.

If the wall is curved, particularly if it is concave toward the sound, the wave fronts *R* of the reflected sound may be curved so as to con-

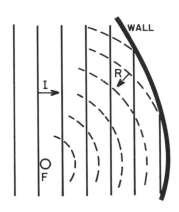

FIG. 16. Sound focus produced by reflection from a concave wall.

centrate the sound in a small region. This is shown in Fig. 16. A region where the sound concentrates in this fashion is called a *focus*; this region is denoted by *F* in Fig. 16.

Reflection of sound is very important to the subject of auditorium acoustics, which we will take up in Ch. 9.

2. *Refraction.* When waves travel through a medium whose properties change slowly with distance, the wave may ultimately be moving in a direction different from that in which it started. This is called *refraction.* Sometimes a condition exists in the atmosphere such that the temperature increases with elevation. Waves from a source on the ground will then travel faster at higher elevations. This produces the refraction effect illustrated in Fig. 17; the wave fronts originally travel-

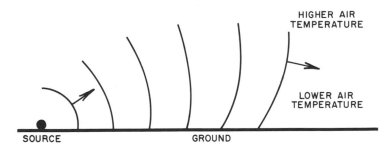

FIG. 17. Refraction of sound waves.

ing away from the source are bent around so as to ultimately head back toward the ground. Refraction is not important to musical acoustics.

3. *Diffraction.* When waves strike an obstacle in the medium, there may be some reflection from it, but in addition the waves which pass

by the obstacle tend to bend around it and fill in the space behind. This ability of waves to go around corners is called *diffraction*. Fig. 18(a) shows wave fronts coming from the left and diffracting around an obstacle; at some distance past it, the wave fronts bending around the two sides of the obstacle have reunited so as to be almost as they would be with the obstacle removed. The phenomenon of diffraction enables us to hear the music at a concert even when some large person sits in front of us and cuts off the view.

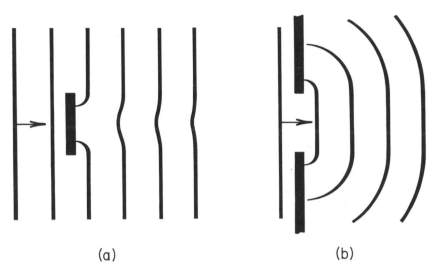

(a)   (b)

FIG. 18. (a) Diffraction of waves around an obstacle. (b) Diffraction of waves through an aperture.

In the same fashion, waves striking an opening in a wall will spread around and fill the space behind it, as shown in Fig. 18(b). This means that we can hear someone talking in the next room if the door is open even though we cannot see him.

4. *Interference.* When waves from two different sources exist simultaneously in the same medium, each wave travels as it would if the other were not there. The displacement of a particular point in the medium cannot have two different values at the same time, however, so when both waves are present, the resulting displacement is the sum of the displacements produced by the individual waves. Since the individual displacements may be positive or negative, the sum may be larger than either, or smaller; the waves may then reinforce one another, or may more or less cancel one another. The result of adding two waves is called *interference*.

If two waves of the same frequency and amplitude proceed from two separate sources, as in Fig. 19(a), there will be regions where

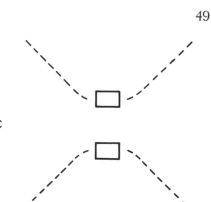

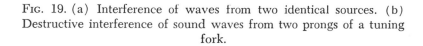

FIG. 19. (a) Interference of waves from two identical sources. (b) Destructive interference of sound waves from two prongs of a tuning fork.

the crests of one set of waves coincide with the crests of the other set, so the resulting amplitude will be twice that of either wave. This is *constructive interference*; the regions where this takes place are shown by the dotted lines marked C in Fig. 19(a). Conversely, there will be other regions where the crests of one set fall on the troughs of the other, and the two will cancel. This is *destructive interference,* and it takes place in the regions D in Fig. 19(a). Along these lines there is no motion of the medium.

A tuning fork provides a good example of destructive interference. The two prongs of the fork oscillate with the same frequency and amplitude, and so provide two identical sound sources. If the fork is sounded, put close to the ear, and rotated slowly about its handle, positions can be found for which no sound is heard. Since the fork is symmetrical about two axes, there must be four such positions. They are shown in Fig. 19(b), which is an end view of a tuning fork; the regions of destructive interference are indicated by the dotted lines.

A particularly important interference effect occurs when two waves of the same frequency and amplitude travel in opposite directions in the same medium; we will discuss this in the next chapter.

Beats

An interference situation of considerable importance to music occurs when two sound sources of slightly different frequencies are sounded together. The resulting sound has a periodic rise and fall in amplitude that is heard as a periodic change in loudness, and which is called a *beat*; the number of times per second this rise and fall occurs is called the *beat frequency*.

The explanation of this effect is simple. Suppose we have two sources Q_1 and Q_2 whose frequencies are 7 cycles per second and 8 cycles per second, respectively. Assume these sources are a considerable distance off to the left in Fig. 20. At a particular instant, the configuration of

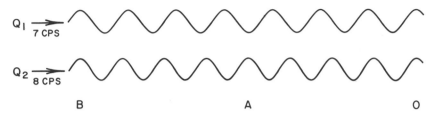

FIG. 20. Production of beats.

the two waves from the sources will be as shown. At that instant, an observer at the point indicated by O will find the wave fronts coinciding, as shown, so that constructive interference exists. The two sets of waves are moving in the direction of the arrows, so one-half second later, the point A will pass the observer. At this instant 4 waves from Q_2 have passed by, but only $3\frac{1}{2}$ from Q_1; the crests of the waves of one set coincide with the troughs of the other so destructive interference exists. Similarly, at the end of one second, point B passes the observer and 8 waves from one set as against 7 from the other have gone by, so constructive interference exists again. The amplitude of the resulting wave alternately increases and decreases, becoming a maximum once each second; this is the beat frequency in this case.

If the two sources had frequencies of 8 cycles per second and 9 cycles per second, we would again find a beat frequency of one cycle per second. The same argument, applied to any two frequencies f_1 and f_2 relatively close together, shows that the beat frequency f_B is given by

$$f_B = f_2 - f_1,\qquad(4)$$

where f_2 is the higher frequency.

Energy and Power in Waves

To create a disturbance in a medium requires the expenditure of energy. Of the work done on the medium by whatever starts the disturbance, a part will go into deforming the medium from its equilibrium shape, and so is momentarily stored as potential energy. The remaining part will be used in setting the parts of the medium into motion, and so appears as kinetic energy. The total energy supplied travels with the disturbance as it moves through the medium.

A source that is producing a continuous train of waves must supply a certain amount of energy to each individual wave. Since a certain number of waves is produced each second, a certain quantity of energy must be supplied each second, so that the production of a continuous train of waves requires a steady expenditure of power by the source.

When a tuning fork is made to vibrate, a certain amount of energy must be given it. In Ch. 2 we saw that the energy supplied the fork slowly disappears because of frictional forces, and the vibrations gradually die away. If the fork is sounded in air, the energy supplied will be used up not only by friction, but also by the production of sound waves, so the vibrations of the fork will decay more rapidly. The process whereby energy is carried away from a source through a medium is called *radiation.*

If we want the tuning fork to vibrate continuously with constant amplitude, we must supply it with a continuous amount of power; this may be done in various ways, as by mechanical bowing, which we will discuss in Ch. 10. Of the power supplied, measured in watts, a certain fraction will go to create sound waves; the power actually appearing as sound is measured in *acoustic watts.* The fraction itself—that is, (acoustic watts)/(watts supplied)—expressed in percent, is called the *efficiency.*

Musical instruments are physical systems which radiate sound waves. Most of them can produce steady sounds of constant amplitude, at least for a time, and so can produce a certain number of watts of acoustic power. This output has been measured for a number of instruments and is shown in Table II.[2] By ordinary electrical standards, the

TABLE II

Measured greatest power outputs
of some musical instruments

	POWER OUTPUT, WATTS
Large orchestra	67
Bass drum	25
Snare drum	12
Cymbals	9.5
Trombone	6.4
Piano	0.44
Trumpet	0.31
Tuba	0.20
Double bass	0.16
Flute	0.055
French horn	0.053
Clarinet	0.050

power output is quite small; that from an entire symphony orchestra would just light one medium-sized light bulb. Note that the percussion supplies a large fraction of the power output of an orchestra.

The efficiency of musical instruments has been measured and turns out to be quite low;[3] the greatest value for any instrument appears to be about one percent. Fortunately, the ear is a very sensitive device, and quite loud sounds can be produced with small amounts of power.

Sound Intensity

Consider a sound source in air that is free to radiate equally well in all directions. A certain amount of power is flowing out from this source. Now imagine a sphere of some given radius surrounding the source, with the source at the center. All of the power radiating from the source must go through the surface of this sphere. There will, then, be a certain amount of power passing through each unit area of the sphere at any point; this quantity, measured in watts per square meter, is called the *sound intensity* at that point.

This is illustrated in Fig. 21. If R is the distance from the sound

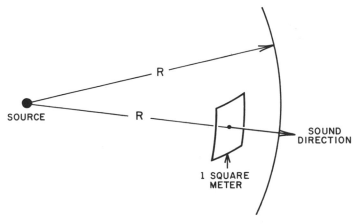

Fig. 21. Sound intensity at a distance from a small source.

source to the point at which we want to know the sound intensity, the area A of the sphere centered at the source and having this radius will be $4\pi R^2$. If P is the acoustic power of the source, the intensity I in watts per square meter at the distance R will be

$$I = \frac{P}{4\pi R^2}. \tag{5}$$

The sound intensity decreases as the distance increases, falling off inversely as the square of the distance. This is just one example of a number of places in physics where this inverse square law applies.

As an example, suppose we calculate the sound intensity at a distance of 3 meters (very nearly 10 feet) from a source radiating one acoustic watt of power. We obtain

$$I = \frac{1}{4\pi(3)^2} = 0.009 \text{ watts/m}^2,$$

or an intensity on the order of 10^{-2} watts per square meter. As we shall see subsequently, this is a very loud sound.

The conditions under which we can use the inverse square law, as in the example above, occur rather seldom. For one thing, most sound sources, including musical instruments, do not radiate uniformly in all directions, but tend to concentrate sound in some directions more than others. For such sources we may still use the definition of sound intensity as the number of watts per square meter passing through an area perpendicular to the direction of the sound, as in Fig. 21. However, we cannot calculate it simply by the use of Eq. (5). Furthermore, this equation is not applicable to a sound source in an enclosure or room, where sound can reflect from the walls.

Theoretically, the sound intensity could be measured with appropriate equipment, but in practice this is difficult. It is much simpler to measure the amplitude of the pressure variation about the ambient pressure; as we saw earlier, this is called the *sound pressure*. For a single wave traveling in one direction, the sound intensity is proportional to the square of the sound pressure.

Sound Absorption

When a sound wave, or any wave, strikes a rigid wall bounding a medium, it is completely reflected. This is because a rigid wall cannot move when a force is applied to it, so it cannot absorb energy from the wave. The intensity of the reflected wave will then be equal to the intensity of the incident wave.

Actual material walls are never completely rigid, however, so they will absorb some energy from sound waves. The intensity I_R of the reflected sound will be somewhat less than the intensity I of the incident sound. The difference between these two must represent the intensity I_A absorbed from the sound, or

$$I_A = I - I_R. \tag{6}$$

For a given material, we may define the *absorption coefficient*, denoted by a, as the ratio of absorbed sound to incident sound, so

$$a = \frac{I_A}{I}. \tag{7}$$

The absorption coefficient is obviously a number that varies between 0 and 1. For $a = 0$, no sound is absorbed, and all is reflected. For $a = 1$, none is reflected, and the sound is totally absorbed.

The value of the absorption coefficient does not depend on the intensity of the incident sound; the same fraction of the sound is absorbed whether the intensity be large or small. The value for a given material does depend somewhat on frequency; most substances absorb sound better at higher frequencies, so the value of the absorption coefficient is greater at higher frequencies.

In Ch. 9, on architectural acoustics, we will give a table of absorption coefficients.

CHAPTER FOUR

COMPLEX
VIBRATIONS
AND RESONANCE

Now that we have discussed wave motions and their transmission through the medium, we are equipped to examine vibrating systems whose behavior is more complicated than those discussed earlier. These more complex systems are the acoustical foundations for the various families of musical instruments.

The Vibrating String: Standing Waves

The simple systems considered earlier had essentially only one vibration frequency. There are actually not very many systems of this kind; most things that can vibrate can do so in more than one way. The string stretched between two supports is an example of such a system, and one that has very important musical applications.

To analyze the behavior of the string fixed at two ends, we make use of an artifice. We first imagine the fixed points to be moved off to infinity, right and left, but with the tension in the string kept the same. (Infinity may be defined as a place so far away that if it is any farther, it doesn't matter.)

We then have the long string already discussed in Ch. 3; let us suppose that a source such as a tuning fork, a considerable distance off on the left, is sending transverse waves down the string toward the right. At a particular instant the wave would look like the top curve of Fig. 1(a). Now suppose a second source, a considerable distance off to the right, is sending a wave to the left along the string; this second source has exactly the same frequency and amplitude as the first, so the two waves traveling in opposite directions have the same amplitude, the same frequency, and hence the same wavelength. This wave would be shown by the second curve in Fig. 1(a) if the first wave were absent. When both waves are present, we will have interference, and the resulting amplitude at any instant is the sum of the amplitudes of each wave at that instant. In Fig. 1(a) the individual traveling waves have been pictured at the instant they would be in phase, with crests and troughs coinciding. The resulting amplitude of the string at that instant is then as shown by the bottom heavy curve in Fig. 1(a), having twice the amplitude of either traveling wave.

Next we may follow the changes in the system by picturing the two traveling waves and the resultant displacement of the string at successive instants during a cycle as shown by Fig. 1(b), (c), (d), and (e). In each representation the positions of the waves traveling to the right and to the left are shown by the first and second curves respectively, and the resultant displacement of the string by the bottom heavy curve.

One-eighth of a cycle from the beginning, as shown by Fig. 1(a), the first wave will have moved one-eighth of a wavelength to the right, and the second wave a similar distance to the left, giving the configuration shown in Fig. 1(b). The resultant displacement of the string is then like that of Fig. 1(a), but with a smaller amplitude.

After another one-eighth cycle—that is, one-quarter cycle from the beginning—the first wave will have moved a total of a quarter of a wavelength to the right, and the second wave the same distance to the left. The two waves will then be exactly out of phase, the peaks of one coinciding with the trough of the other, as shown in Fig. 1(c). The resulting displacement at this instant is then zero everywhere along the string.

Another one-eighth period later (three-eighths of a cycle from the beginning), each traveling wave will have moved to the positions shown in Fig. 1(d). The resulting displacement of the string is then like that of Fig. 1(b) but in the opposite sense, so that positive displacements in Fig. 1(b) are now negative, and conversely.

Another one-eighth period later (one-half cycle from the beginning) each wave will have moved to the positions shown in Fig. 1(e). They are now back in phase, and the resultant displacement is now as it was in Fig. 1(a), but again in the opposite sense.

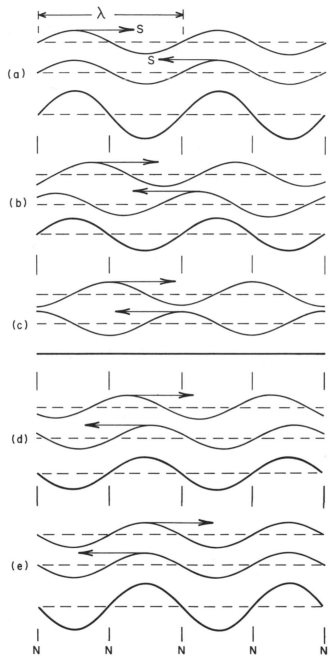

FIG. 1. Production of a standing wave in a string. Diagrams (a), (b), (c), (d), and (e) are one-eighth of a cycle apart.

At succeeding one-eighth cycle increments, the same considerations will show that the string retraces the positions (d), (c), (b), and (a) of Fig. 1. At this time it is back in its original position, and the whole cycle starts over.

The result is a motion of the string up and down in a series of loops as shown in Fig. 2, where the solid curve shows the maximum displacement in one direction, and the dashed curve shows the displacement one-half cycle later. This kind of motion is called a *standing wave,* since the wave is not moving in either direction along the string. Practically all musical instruments generate standing waves, as we will see.

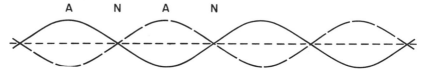

FIG. 2. Standing wave on a long string.

In a standing wave, there are certain points along the string that do not move; these points are designated by N in Figs. 1 and 2, and are called *nodes*. The vibrating segment of string between two nodes is a *loop*. Halfway between any two nodes is a point where the vibration has the greatest amplitude; this point is called an *antinode*. Points marked A in Fig. 2 indicate positions of antinodes.

Referring back to Fig. 1(a), we see an important relation: the distance between two adjacent nodes of a standing wave is equal to one-half wavelength of the two traveling waves that form the standing wave.

With this information we are able to work out the oscillation frequencies of a string fixed at both ends. Suppose its length is L meters. In the long string, the wavelength of the traveling waves depends on the frequency of the sources by the fundamental wave relation given in Eq. (2) of Ch. 3, namely $S = f\lambda$. The distance between two adjacent nodes will then also depend on the frequency. Suppose it to be adjusted until the distance between nodes is L. Since the nodes are points on the string that are not moving, we can clamp the string at these points without affecting the vibration of the string in between. The string outside these clamps could then be cut off and removed, leaving the string clamped at its two ends still vibrating, as in Fig. 3(a).

The frequency of this vibration may now be obtained very simply. The two fixed ends of the string are nodes, and since the distance between nodes is half the wavelength λ_1 of the traveling waves producing this standing wave, we have

$$\lambda_1 = 2L. \tag{1}$$

FIG. 3. First four vibration modes of a string fastened at both ends.

Now from the fundamental wave relation we have $f_1 = S/\lambda_1$, where f_1 is the vibration frequency producing the wavelength λ_1. Hence we obtain

$$f_1 = \frac{S}{2L} \tag{2}$$

as the frequency for the vibration shown in Fig. 2(a). This is the frequency for which we can just fit one loop of the standing wave into the space between the clamps. This particular vibration is called the *fundamental mode* of the string. Since the speed S of waves on the string is given by Eq. (1) in Ch. 3, we may write this

$$f_1 = \frac{1}{2L}\sqrt{\frac{T}{\rho}}. \tag{3}$$

This, then, is the fundamental vibration frequency of a string of length L, fixed at both ends, having a mass ρ kilograms per meter length and stretched with a tension T newtons.

This is not the only way the string can vibrate. Suppose we increase the frequency of oscillation of the long string and thus shorten the loops until we can fit two of them into the space between supports, as in Fig. 3(b). The new wavelength is now given by

$$\lambda_2 = L, \tag{4}$$

so the oscillation frequency is

$$f_2 = \frac{S}{L} = 2\frac{S}{2L} = 2f_1. \tag{5}$$

This is the next vibration mode of the string, and it has a frequency exactly twice the fundamental.

Similarly, we may increase the frequency until three loops of a standing wave will fit the string. From Fig. 3(c) we have

$$\lambda_3 = \frac{2}{3}L, \tag{6}$$

so the frequency of this mode is

$$f_3 = 3\frac{S}{2L} = 3f_1, \tag{7}$$

or exactly three times the fundamental. Higher frequency modes will be found in the same way.

The string stretched between supports can thus vibrate in a number of modes that are related quite simply. The frequency of the lowest or fundamental modes is given by Eq. (3) above. The other modes have frequencies that are *integral multiples*—that is, 2,3,4, and so on times the fundamental frequency.

A medium can carry many waves simultaneously, as we stated earlier, so the stretched string can vibrate in many modes simultaneously. If we set the string into vibration by plucking it—that is, by drawing it to one side and suddenly releasing it—all the modes will be present, with amplitudes that depend on the point along the string at which it was plucked. The existence of these modes may be shown quite easily. If we touch the vibrating string at some point, the vibrations will be stopped for all except those modes that have a node at the point touched; these can keep on vibrating. Hence if the string is plucked at some random point and then touched at the center, it will be found that the second mode is vibrating. (This may be determined by noting that the pitch of the emitted sound rises an octave, so that the frequency is twice as great; we will discuss this in Ch. 8.) Similarly, if we touch the vibrating string at one-third the distance from one end, the third mode will be found to be vibrating. (The pitch rises a twelfth.) In general, then, the stretched string vibrates in a complex fashion with many modes and frequencies present simultaneously.

If the string is plucked at a point and then touched at that same point, all vibrations stop completely. If this point were a node for a particular mode, however, that mode could go on vibrating. The fact that it does not do so, however, means that if the string is plucked at a point that is a node for a particular mode of vibration, that mode is not produced.

Vibrating Air Columns

The long air column will transmit longitudinal waves, just as the long string will transmit transverse waves. If we create two waves of equal amplitude and frequency moving in opposite directions in the column, we will obtain a longitudinal standing wave in it. The argument is similar to that for the string, so we need not repeat it. Fig. 4(a) shows

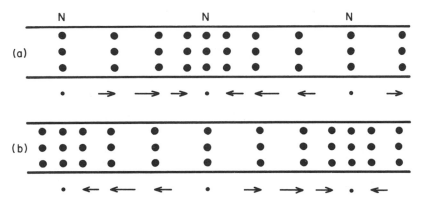

FIG. 4. Longitudinal standing wave in an air column. (a) At an instant of maximum displacement of the air molecules. (b) One-half cycle later.

the configurations of a few chosen air molecules in the column for the instant at which their displacements are at a maximum, and Fig. 4(b) shows the configuration one-half cycle later; these figures correspond to the same parts of the cycle as Fig. 1(a) and (e) respectively. The points marked N are the nodes of the standing wave, being places where the air does not oscillate.

The graphic representation of Ch. 3 may be used to advantage here. We plot the displacements as given by the arrows under Fig. 4(a) and (b) against distance along the column. We then get Fig. 5, where the

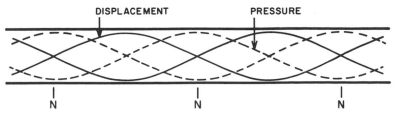

FIG. 5. Graphic representation of a longitudinal standing wave.

separation of the solid curves is a measure of the amplitude of vibration of the air molecules. The displacement nodes are marked N as before.

In Fig. 4(a) we see that the molecules are pulled apart at the first node on the left, creating a rarefaction, or a region of lower pressure; conversely, they are crowded together at the next node, creating a condensation, or region of higher pressure. In Fig. 4(b) the picture is reversed. We thus have a standing wave of pressure in the air column. This is plotted as the dotted pair of curves in Fig. 5. We see that the displacement nodes correspond to the pressure antinodes, and conversely.

We may now work out the vibration frequencies of air columns of given lengths. Since the nodes in Figs. 4 and 5 are places where the air is not in motion, we could put solid diaphragms across the tube at these points without disturbing the vibrations elsewhere. If we now cut the tube at the diaphragms and throw away all but the piece between them, we will have an air column closed at both ends with a standing wave inside it, as shown in Fig. 6(a). This system corresponds

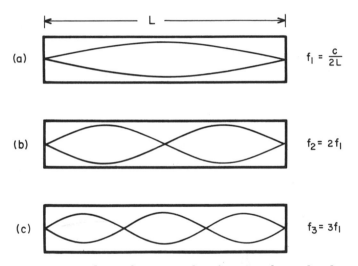

FIG. 6. First three vibration modes of an air column closed at both ends.

precisely to the string whose modes are shown in Fig. 3, having nodes at the two ends. Hence we do not need to repeat arguments, and can say that the tube closed at both ends has modes of vibration like those of the string. The first of these obviously has a fundamental frequency

$$f_1 = \frac{c}{2L},\tag{8}$$

where L is the length of the tube and c is the speed of sound in the tube. The remaining modes will have frequencies that are the integral multiples 2,3,4, and so on times the fundamental frequency.

This tube is of no musical importance, since no sound can get out of it, so we will not refer to it further. Much more important is the tube open at both ends, which we will from now on refer to as an *open tube*. At an open end the pressure must be simply the ambient air pressure, so the amplitude of the pressure vibration must be very nearly zero. The open end is hence a pressure node, so it must be a displacement antinode. We can therefore fit a standing wave into this tube by putting displacement antinodes at the open ends, as shown in Fig. 7(a). Since the distance between two antinodes is the same as between two nodes, all the calculations are the same as previously, so we find that the frequency of the fundamental mode is again given by Eq. (8) above. Higher modes will fit as shown in Fig. 7(b) and (c), giving frequencies the same as those in Fig. 6(b) and (c). Hence the open tube also has modes whose frequencies are integral multiples of the fundamental frequency.

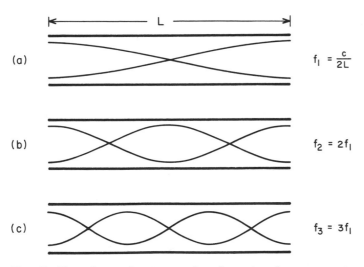

(a) $f_1 = \dfrac{c}{2L}$

(b) $f_2 = 2f_1$

(c) $f_3 = 3f_1$

FIG. 7. First three vibration modes of an air column open at both ends.

A tube open at one end and closed at the other, which we shall refer to as a *closed tube*, is somewhat more complicated. The closed end must be a displacement node and the open end an antinode. The longest standing wave that will fit into this tube is then as shown in Fig. 8(a). The distance between a node and the adjacent antinode is one-

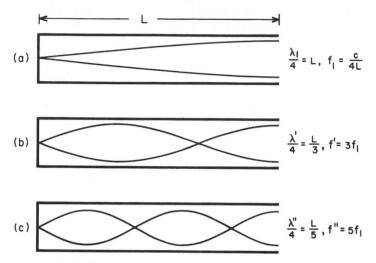

FIG. 8. First three vibration modes of an air column open at one end and closed at the other.

quarter of the wavelength of the traveling waves. This gives the wavelength $\lambda_1 = 4L$, so the fundamental frequency f_1 is

$$f_1 = \frac{c}{4L}, \tag{9}$$

which is half as large as that for the open tube.

The next mode that will fit the tube so as to give a node at the closed end and an antinode at the open end is shown in Fig. 8(b). One loop of the standing wave now occupies two-thirds the length of the tube. This gives for its wavelength λ' the relation

$$\frac{\lambda'}{2} = \frac{2}{3}L, \tag{10}$$

so the frequency for this mode is

$$f' = 3\frac{c}{4L} = 3f_1. \tag{11}$$

Similarly, the next higher mode above this will be as shown in Fig. 8(c) with a wavelength λ'' given by

$$\frac{\lambda''}{2} = \frac{2}{5}L; \tag{12}$$

the frequency of this mode is then

$$f'' = 5\frac{c}{4L} = 5f_1. \tag{13}$$

Higher modes will be found similarly.

The mode frequencies for the closed tube are thus the odd integral multiples 1,3,5,7,9, and so on times the fundamental frequency; all the even multiples are missing.

At the open end of an air column containing a standing wave, the air is moving in and out of the open end, and its motion actually extends a little way past the end. This makes the tube appear longer than it actually is by an amount called the *end correction*. For a cylindrical pipe of radius r, the end correction has been calculated to be $0.61r$.[1] This amount should be added to the actual length of the pipe for each open end to get the proper value of effective length L to use in the formulas above. The end correction varies with frequency, but the variation is too small to be of any practical significance.

There is another shape of air column which is of musical importance; this is the cone. The mathematics necessary to work out the vibration frequencies of the modes of a cone is too complicated to give here. It shows that the fundamental frequency is $c/2L$ where L is the length of the cone, and that the higher modes have frequencies that are again the integral multiples 2,3,4, and so on times the fundamental frequency. The cone thus behaves like the open tube of the same length.

There is one other system that behaves like the air column; this is a metal bar vibrating longitudinally. If the two ends of the bar are free, they must be displacement antinodes, so the vibrations of the bar will be like those of Fig. 7 for the open air column. The same formulas for vibration frequencies will apply, so the fundamental frequency of the bar will be

$$f_1 = \frac{S}{2L}, \tag{14}$$

where S is the speed of sound in the metal of the bar. If the bar is clamped at the center, this point must be a node, and only the odd multiples of the fundamental frequency will be possible.

Other Vibrating Systems

It must not be thought that all oscillating systems that have more than one mode of vibration will follow the simple relationships displayed by the string and the air column. These simple systems are exceptions; for most systems having many modes, the vibration frequencies are related in more complicated ways. For example, if we drill a hole in the side of a tube, or close off the tip of a cone, producing distorted air columns as in Fig. 9, we will change the frequencies of the modes so that they are no longer simple integral multiples of the fundamental frequency.

Another example is provided by the transverse vibrations of a metal bar that is bent slightly out of shape and released. These are much more complicated than the longitudinal vibrations mentioned above.

FIG. 9. Distorted air columns.

The restoring forces that appear when the bar is distorted laterally are due to the elasticity of the material, which resists bending. This makes the analysis quite complicated; for example, the speed of a wave along a stiff bar is not independent of frequency, as it is for the string and the air column. For a bar clamped at one end and free at the other, the first three modes are as shown in Fig. 10. (The amplitudes of the vibrations have been drawn large for clarity; they would be much smaller for actual vibrations.) If f_1 is the frequency of the lowest mode, shown in Fig. 10(a), the next two modes have frequencies given by $6.27f_1$ and $17.55f_1$. The frequency f_1 depends on the thickness of the bar and the density and elasticity of the material.

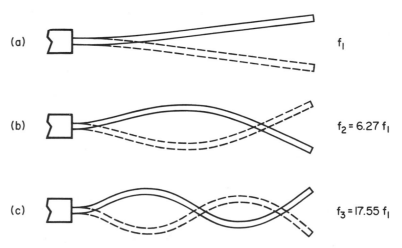

FIG. 10. First three transverse vibration modes of a metal bar clamped at one end.

The tuning fork is actually two bars joined together at one end, and when oscillating in its normal manner the two bars (prongs) each vibrate as in Fig. 10(a) with approximately the frequency f_1. The vibration of the whole fork is then as shown (somewhat exaggerated, for clarity) in Fig. 11(a), with the two extreme displacements shown by solid and dotted lines respectively. In their motion the bars pivot about the nodes marked X in the figure, causing the handle to oscillate in the direction shown by the arrow. If the handle is pressed against another object, its vibrations will then be communicated to that object.

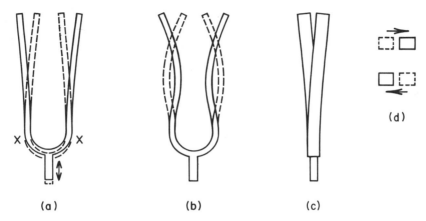

(a)   (b)   (c)

Fig. 11. Vibrations of a tuning fork. (a) Normal vibration. (b) "Clang" tone. (c) Another mode of vibration. (d) End view of this mode.

The tuning fork bars can also vibrate in higher modes. The next mode will be as shown in Fig. 11(b), and will have a frequency some six times that of the normal mode. When a tuning fork is struck with a hammer, this and still higher modes are produced, giving a high-pitched sound called the *clang-tone*. It usually dies away rather rapidly, leaving the fork to vibrate in its normal mode.

The fork has another mode that is generally overlooked. If the handle is clamped so that the fork cannot rotate, the prongs can vibrate as shown in Fig. 11(c), with an end view shown in Fig. 11(d). The frequency of this mode is not necessarily high and may be fairly close to that of the normal vibration.

We will discuss other modes of vibration of single bars in connection with percussion instruments.

A *membrane* is a thin sheet of material that can be put under tension. If a membrane is stretched over and fastened to a frame, it can vibrate in a number of ways. The lowest frequency mode will be that for which the membrane vibrates all over, and will depend on the size of the frame, the density of the membrane, and the tension in it. Higher modes can also exist; for these there will be lines called *nodal lines,* along which there is no motion. These lines can be demonstrated by sprinkling fine sand on the vibrating membrane; the sand is thrown away from the vibrating portions and piles up along the nodal lines.

For a rectangular membrane stretched equally in all directions, some of the vibration modes are as shown in Fig. 12. The nodal lines are

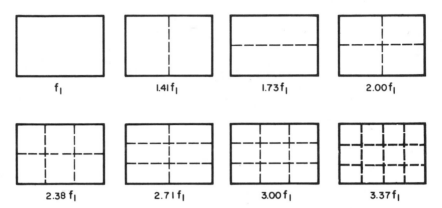

FIG. 12. Some of the modes of vibration of a stretched rectangular membrane. The length of the membrane is 1.41 times the width.

shown dotted; as with the vibrating string, adjacent segments are always moving in opposite directions. The frequencies of these modes are given in terms of the fundamental frequency. There are modes for which the frequencies are 2,3,4, and so on times the fundamental, but there are many others not so simply related.

The circular membrane is important musically; we will discuss it in Ch. 14, in connection with the percussion instruments.

Plates are even more complicated than bars and membranes, having many modes of vibration spaced quite close together in frequency. Some of the modes of a square plate clamped at the center (which must then be a node) and bowed at some point along the edge are shown in Fig. 13. The nodal lines may be shown by sprinkling sand on

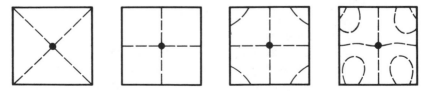

FIG. 13. Some of the modes of vibration of a square plate fixed at the center and set into oscillation by bowing.

the plate, as above. A rectangular plate will have even more modes than a square plate, as will a plate of irregular shape such as the back or top of a violin.

Sound Radiation from Vibrating Systems

All of the vibrating systems we have discussed can communicate vibrations to a certain degree to the surrounding air and hence radiate sound. Some systems do this much better than others, however. The vibrating prongs of a tuning fork, for example, produce very little sound, since they move back and forth through the air without particularly disturbing it. As a result, the tuning fork cannot be heard more than a few inches away. However, if the handle of the vibrating fork is pressed against a flat surface such as a table top, the sound may be heard throughout the room. The handle of the fork causes the table top to vibrate, as already explained above, and the large vibrating areas of the table radiate sound much better than the fork itself.

A vibrating string also produces very little sound by itself, since it is too narrow to disturb very much air. To produce sound from the string, it is necessary to furnish some better means of communicating its vibrations to the air. This can be done as shown in Fig. 14. The string passes over a support which rests on a wide plate, usually wood, sometimes called the *soundboard;* string vibrations are communicated to the soundboard, which then radiates to the surrounding air.

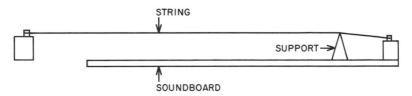

FIG. 14. String and soundboard.

Air columns, on the other hand, do not need assistance to radiate sound. The pressure amplitude just outside the open end of a vibrating air column is very nearly but not quite zero, and this oscillating pressure causes sound to radiate from the open end. The amount of energy so radiated each cycle is a quite small fraction of the energy possessed by the vibrating air column, but it is enough for practical musical purposes. We note from Eq. (8) above that the open tube vibrating in its fundamental mode has a length just half the wavelength of the sound it radiates; likewise, from Eq. (9), we see that the closed tube vibrating in its fundamental mode is a quarter wavelength long.

If a tuning fork is mounted on a hollow wooden box, as shown in Fig. 15, it will cause the top of the box to vibrate. If the box is open at one end, its interior is an air column which has a certain vibration frequency. If this frequency is the same as that of the fork, then the vibrations of the fork will build up vibrations of considerable amplitude in the air column, which will then radiate sound from the open end of the box. Tuning forks are frequently mounted in this way to increase their sound output.

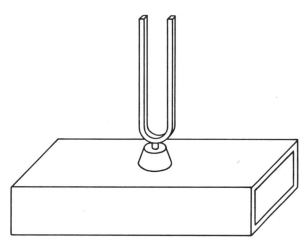

FIG. 15. Tuning fork mounted on resonating box.

Resonance

The combination of tuning fork and box illustrates a situation of great importance in physics. Whenever a system that can vibrate with a certain frequency is acted upon from the outside by a periodic disturbance that has the same frequency, vibrations of large amplitude can be produced in the system. This situation is called *resonance*. If the frequency of the external disturbances is either somewhat lower or somewhat higher than that of the system, the vibration amplitude built up in the system is not as large as it is when the two are the same. The periodic external disturbance is called the *excitation*. The frequency of the system is very often called the *resonance frequency*, since it is the frequency for which excitation will produce the greatest response. Similarly, for a system with more than one mode of vibration, we may call the frequencies of these modes the *resonance frequencies*, and the modes themselves may be referred to as *resonance modes*, or simply as *resonances*. We will use these terms very often in our discussions of musical instruments.

There are many examples of resonance. In Ch. 2 we discussed the vibration of the mass on the spring when the mass was given a small push each cycle so as to add a little energy to the system. The effect of this was to build up a large amplitude oscillation in the system; this was therefore a resonance condition. The effect of excitation is to add a small amount of energy to the system at each cycle until the vibrations build up to an amplitude such that the same amount of energy is lost each cycle.

One of the simplest demonstrations of resonance is that furnished by a *pendulum*, which is simply a small mass hung on a string. The pendulum has a frequency that depends only on the length of the string. If the string is held in the hand and moved back and forth over a small range at the frequency of the pendulum, oscillations of large amplitude can be built up.

Two identical tuning forks mounted on resonating boxes for better sound radiation also furnish a good demonstration of resonance. If one is set vibrating and then stopped, the other will be found to be vibrating. The sound radiated by the first fork provides the excitation for the second. This effect can take place even when the forks are several feet apart. Similarly, a sound of the right frequency from some source such as a loudspeaker (to be discussed in Ch. 15) can cause the fork to vibrate. If the frequency of the sound is different from that of the fork by a small amount (a few tenths of a percent) the effect disappears.

A simple laboratory experiment demonstrating resonance is shown in Fig. 16. A glass tube contains water whose level may be changed by adjusting the height of a reservoir connected to the tube by means of a flexible hose. This allows the length of the air column in the tube to be easily altered. If a tuning fork (not on a radiating box) is made to vibrate and then held above the open end, a length of air column can be found where the normally inaudible sound of the fork becomes quite loud. This length is that for which the vibration frequency of the air column is the same as that of the fork, so the column is set into oscillation by resonance. It is the air column that is radiating the sound heard, not the fork.

Helmholtz Resonator

There is one acoustical vibrating system that behaves like the mass on the spring that we discussed in Ch. 2. However, it uses air both to supply mass and to furnish restoring force, so we have deferred its description until we had discussed air as a medium.

Suppose we have an enclosure made of some reasonably rigid material. Its exact shape does not matter as long as it is not too long and

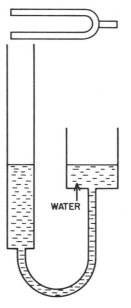

FIG. 16. Experiment demonstrating resonance.

narrow. The interior of the enclosure communicates to the outside through a relatively narrow neck whose shape is not important. The structure will be something like that shown in Fig. 17(a): an empty bottle with not too wide a mouth will provide such a system.

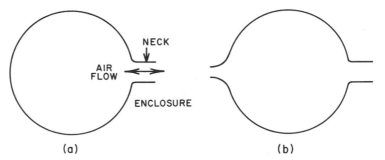

FIG. 17. (a) Helmholtz resonator. (b) Resonator adapted for hearing a sound of specific frequency in a mixture of frequencies.

Now assume that we put a cork in the neck to close it off and then pump some air into the interior of the enclosure through a hole in the cork—like pumping a tire. The pressure inside the enclosure will then become higher than that outside, in proportion to the amount of air pumped in. After we do this, suppose we suddenly remove the cork. The excess pressure inside the enclosure then pushes on the mass of air in the neck, causing it to accelerate and flow out. As it does so, the pressure inside falls until it becomes the same as the outside ambient pressure. However, at this instant the mass of air in the neck has acquired a certain momentum, so it keeps on moving out of the enclosure. This reduces the pressure inside below that outside; the flow of air through the neck then slows down and finally stops. At this instant the pressure inside is as far below that outside as it was above at the start, when the cork was removed. Hence air starts flowing into the enclosure again and the same process occurs in the reverse direction. The pressure inside then rises again to its initial value, and the cycle repeats.

The result is a vibration of air in and out of the neck of the enclosure, exactly like that of the mass on the spring discussed in Ch. 2. This arrangement is called a *Helmholtz resonator,* after the famous German physicist who developed it. We shall refer to it occasionally in subsequent discussions of the behavior of musical instruments.

The restoring force in the resonator is provided by the excess pressure inside the enclosure when it contains excess air; the larger the volume of the enclosure, the more excess air it can contain for a given excess pressure. This corresponds to a spring with a small stiffness, which will give a large displacement for a given force. Consequently, the frequency of vibration of the resonator will be lowered if the volume of the enclosure is made larger. The frequency also depends on the *acoustic mass* of the air in the neck. This quantity depends on the length of the neck; increasing this length will increase the actual mass of air, and so lower the vibration frequency. The longer neck thus corresponds to a larger acoustic mass. However, increasing the area of the neck will allow air to flow out faster for a given excess pressure, and so will increase the frequency; this corresponds to a smaller acoustic mass. The acoustic mass of the system is then inversely proportional to the area of the neck.

The Helmholtz resonator is so named because it makes use of resonance to detect sounds of a particular frequency in the presence of sounds of other frequencies. For this purpose, the system is altered by providing another short open neck on the enclosure, as shown in Fig. 17(b). This neck is then closed off again by inserting it into the ear. An external sound whose frequency is close enough to that of the resonator will then excite strong vibrations in it, and the ear will hear these as a loud tone. Hence the resonator essentially amplifies those sounds of a given frequency that are part of a complex mixture of sounds, enabling them to be distinguished. As a means of analyzing sounds, the Helmholtz resonator is quite obsolete; much more sensitive electronic devices have superseded it.

Resonance Curves

Suppose now the small neck (on the left) of the Helmholtz resonator is closed off by attaching to it a small microphone. This is a device which will produce an electrical signal when the pressure on it changes; we will discuss it further in Ch. 15. With it we can measure the amplitude of the pressure oscillation in the bulb of the resonator. Now we provide excitation by producing a sound outside the open neck of the resonator from some external source such as a loudspeaker. The frequency of this external sound may be varied over any range we please. The amplitude of the external sound pressure at the open neck is kept constant as the frequency is varied, being checked by means of another microphone (not shown) placed near the open neck.

We now measure the amplitude of the pressure vibration in the bulb as the frequency of the external excitation is varied, and plot a graph of the results. What we get is the solid curve shown in Fig. 18; this is

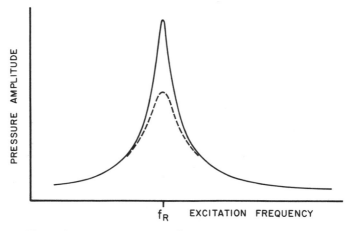

FIG. 18. Resonance curve for a Helmholtz resonator.

called a *resonance curve*. It shows that the pressure vibration in the bulb increases in amplitude as the excitation frequency rises, reaches a sharp maximum at the resonance frequency, and then falls. We will subsequently refer to this method of obtaining resonance curves as the *external excitation* method.

If the inside of the open neck is roughened so that there is more friction acting on the air oscillating in the neck, the vibrations in the bulb will not build up to as large an amplitude at resonance because of the increased frictional energy losses. The resonance frequency, however, will be essentially unchanged, and the resonance curve will look like the dotted curve of Fig. 18.

Resonance curves of other systems may be measured by this method. By installing a microphone in the closed end of a tube and exciting the open end as above, we may obtain the resonance curve of such a closed tube. The curve for a piece of metal tube of 1.6 centimeters (⅝ inches) internal diameter and 58 centimeters (23 inches) long is shown in Fig. 19. The relative frequency scale along the bottom, indicated by short vertical lines below the horizontal axis, is expressed in terms of the fundamental frequency shown which is designated as relative frequency 1. The positions of the resonance peaks are indicated along the horizontal axis by short vertical lines above the axis.

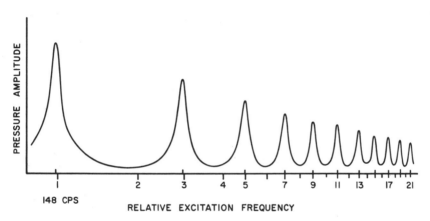

Fig. 19. Resonance curve for a metal tube closed at one end.

We see in Fig. 19 that the resonance frequencies are all integral odd multiples of the fundamental frequency, as we worked out earlier. The slight discrepancies between the positions of the resonance frequencies and the positions of frequencies 1 and 3 are due to friction between the oscillating air column and the walls of the tube. This friction slows down the vibrations somewhat, and has the largest effect at low frequencies.

In Fig. 19, note that the frequency scale along the bottom of the chart is not uniform. Instead, it is a *logarithmic frequency scale*, so called because the distance along the horizontal axis is proportional to the logarithm of the frequency. (See the Appendix for a discussion of logarithms. The term *scale* as used here has no musical significance.) A scale for which the distance along the axis is proportional to frequency, instead of its logarithm, is called a *linear frequency scale*.

The logarithmic scale is very useful in acoustical work. This is because, in acoustics as well as in music, the octave (to be discussed in Ch. 8) is an interval of fundamental importance. In Fig. 20 we see a

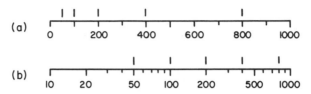

FIG. 20. (a) Linear frequency scale. (b) Logarithmic frequency scale. Vertical lines show frequencies an octave apart, starting arbitrarily at 50 cycles per second.

linear frequency scale and a logarithmic frequency scale. Some frequencies an octave apart are shown by vertical lines on the two scales, starting with 50 cycles per second. It will be seen that the logarithmic scale gives a uniform spacing for these frequencies; this is a great convenience. The piano keyboard is approximately a logarithmic frequency scale; it is hoped that this revelation will not impel pianists to change to some other instrument.

Resonance curves such as that of Fig. 19 will be very helpful later on when we discuss musical instruments.

PART II

THE RECEPTION OF MUSICAL SOUNDS

CHAPTER FIVE

THE EAR: INTENSITY AND LOUDNESS LEVELS

We have now developed the physical and acoustical foundations for a discussion of the acoustics of music, so that eventually we may examine the physical behavior of musical instruments as sources of sounds. However, we cannot do this properly until we have learned a little about the listener's reaction to these sounds, and this involves the behavior of the receptor that we call the *ear*. This organ is a vital part of a musician's equipment; it is as complicated and as little understood as any of his instruments.

Structure of the Ear

The actual appearance, structure, and relationship of the various parts of the ear are difficult to show in a literal reproduction; such drawings are available in the literature.[1] For our purposes, a schematic diagram will show better the relationship and functioning of the various parts of the ear; this is given in Fig. 1.

For descriptive purposes, the ear is divided into three parts: the outer, the middle, and the inner ear. The *outer ear* begins with the obvious external part called the *pinna*, which serves to concentrate sound waves somewhat into the opening of the auditory canal. In most animals (and some people) the pinna is moveable to help collect sound,

The Reception of Musical Sounds

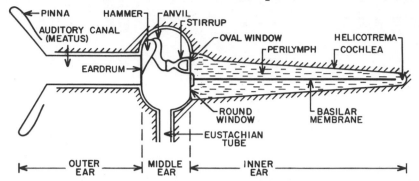

Fɪɢ. 1. Schematic diagram of the human ear, with the cochlea uncoiled.

but the human pinna has actually become rather useless for this purpose. The *auditory canal* (also called the *meatus*) is a tube about three centimeters long, which is closed at the inner end by a membrane called the *eardrum*.

On the inside of the eardrum is the region of the *middle ear*. It is connected to the back of the throat by the *Eustachian tube* so that changes in the ambient air pressure may be equalized and not cause large distortions of the eardrum. The middle ear contains a chain of three small bones—the *hammer,* the *anvil,* and the *stirrup*—which are connected to each other as shown; these connections are not rigid. The hammer is connected to the eardrum so as to move with it. At the other end of the chain, the stirrup is attached to a membrane called the *oval window.*

The *inner ear* is that portion inside the oval window. It consists in part of a cavity in the bony structure of the skull called the *cochlea,* which resembles a snail shell in general shape, having two and three-quarter turns overall. For convenience, this structure has been shown straightened out in Fig. 1. It is filled with a fluid (the *perilymph*) and is divided down its length by a soft partition called the *basilar membrane.* This forms two long chambers, one of which is separated from the middle ear by the oval window attached to the stirrup. The other chamber is separated from the middle ear by the *round window.* The two chambers are connected together at the far end by a small opening in the basilar membrane called the *helicotrema.* Many thousand nerve endings embedded in the basilar membrane sense disturbances in it and transmit this information to the brain. Attached to the cochlea (but not shown in Fig. 1) are the *semicircular canals.* They play no part in hearing, but orient us as to up and down and so are essential in maintaining equilibrium.

Sound waves from the outside pass down the auditory canal to the eardrum, exerting pressure on it and causing it to vibrate. This vibration is transmitted through the chain of three bones to the oval window of the cochlea. The arrangement of the chain of bones is equivalent to a lever system such that the motion of the oval window is about one-third that of the eardrum. The increase in force provided by this lever system, together with the small area of the oval window compared to that of the eardrum, can result in a pressure on the oval window of some 10 to 20 times the pressure on the eardrum.[2] The chain of bones is also arranged so that for very loud sounds the motion of the oval window compared to that of the eardrum is reduced still further; this lessens the likelihood of ear damage from loud sounds.[3]

The pressure on the oval window is transmitted to the adjacent fluid inside the cochlea. Fluids are quite incompressible, and the bony structure in which the cochlea is imbedded is quite unyielding. Hence, to allow the fluid to move when pressure is applied to the oval window, the round window is provided. This is an opening covered with a membrane which can yield under pressure. A motion of the oval window inward is accompanied by a motion of the round window outward, together with a motion of the cochlear fluid. The fluid motions in turn produce motions in the basilar membrane; these are sensed by the nerve endings embedded in the membrane and the resulting stimuli transmitted to the brain.

A great deal of work has been done to determine the manner in which the basilar membrane vibrates when sound waves are received by the ear. It appears that when the oval window is vibrated at a single frequency, waves travel along the basilar membrane away from the window.[4] As each individual wave moves, its amplitude increases until the wave reaches a certain point along the membrane, after which its amplitude quickly diminishes. The point along the membrane at which the greatest amplitude occurs depends on the frequency of vibration; for high frequencies it is close to the oval window, and for low frequencies it is at the far end of the membrane, near the helicotrema. The sense of pitch (related to vibration frequency) is thus partly determined by the place along the basilar membrane where the vibration amplitude is largest. There must be other factors also, since for sounds close together in frequency, especially at low frequencies, the difference in motion of the basilar membrane does not appear great enough to account for the pitch discrimination of a good musician. The behavior of the inner ear together with its nerve connections to the brain is still quite imperfectly understood and is the subject of a great deal of investigation.

Sound Intensity and Hearing

The ear is an extremely sensitive device. To see how sensitive, let us do an experiment with sound intensities. In Ch. 3 we calculated that a sound source of one watt would produce a sound intensity of 10^{-2} watts per square meter at a distance of approximately ten feet. (This distance from a sound source in the open will give intensities comparable to those produced by the same source in an average sort of rehearsal hall, as we will show in Ch. 9.) Since the response of the ear to sounds depends on frequency, let us choose as a starting point the rather irritating frequency of 1000 cycles per second, since this is a standard frequency for many audio measurements.

If we now listen to this sound intensity at this frequency, we would judge it to be a very loud sound, rather uncomfortably so at this frequency. The musician would call it *fff*, yet it is produced by a source whose power, by ordinary everyday standards, is quite small. Now let us reduce the intensity by a factor of 100, either by moving the source to a distance of 100 feet or by leaving it where it is and reducing its power to 0.01 watt. The intensity is now 10^{-4} watts per square meter, but the sound still appears to be quite loud—on the order of *forte*. Another factor of 100 will bring the intensity down to 10^{-6} watts per square meter, and now the sound is fairly soft—on the order of *piano*. Another factor of 100 will get the intensity down to 10^{-8} watts per square meter, where we would judge it to be very soft—on the order of *ppp*. Our useful musical range of sound intensities is thus about a million to one.

The intensity of 10^{-8} watts per square meter is not the lowest value that can be heard; it is rather the intensity for which the sound is likely to be obscured by extraneous noises. If we exclude all outside sounds, we can push the intensity down to the point where the sound can just barely be heard. This is called the *threshold intensity;* it is of the order of 10^{-12} watts per square meter at 1000 cycles per second. The exact value varies with the individual listener and is somewhat more than this, but this figure has been chosen as a standard of reference.[5]

In the other direction, toward very loud sounds, it is a matter of what the ear can endure. If we start from the figure 10^{-2} watts per square meter and increase the intensity, we find that above about one watt per square meter sound produces actual physiological pain. This value may be taken as a necessary upper limit to sound intensity.

Intensity Level: Decibels

In the discussion above we related the musical description of a sound sensation to the intensity of the sound in watts per square meter. This is not very practical, especially for musicians; the statement "10^{-4} watts per square meter" would not mean much to most people. We could use the threshold intensity as a starting point, and correlate the sound sensation with the factor by which the given sound is louder than the threshold; that is, if I is the sound intensity and I_0 is the threshold intensity, we would use the ratio I/I_0 as a measure. For the *ppp* sound of 10^{-8} watts per square meter this ratio would be 10^4; that is, this sound would be ten thousand times the threshold intensity. Similarly, the *fff* sound would be 10^{10} or ten billion times the threshold intensity. This is obviously a rather cumbersome representation also.

In the above numbers, the important quantity is the power of ten that appears. This suggests that we use logarithms. (These are worked out in the Appendix.) A very useful measure of a sound is then expressed in terms of its *intensity level*. If we denote this by L, it is defined as follows:

$$L = 10 \ \log \frac{I}{I_0}. \tag{1}$$

The unit of intensity level is the *decibel*, abbreviated dB.

From this definition we obtain quite a useful scale with which to measure sounds. If we apply Eq. (1) to our earlier examples, we get the results shown in Table I. This gives the level of the threshold of

TABLE I
Sound intensities expressed in decibels

	INTENSITY, w/m^2	RATIO I/I_0	LEVEL, dB
Threshold			
of pain	$10^0 = 1$	10^{12}	120
fff	10^{-2}	10^{10}	100
f	10^{-4}	10^8	80
p	10^{-6}	10^6	60
ppp	10^{-8}	10^4	40
Threshold			
of hearing	10^{-12}	1	0

hearing as zero decibels, which seems sensible. The range from *ppp* to *fff* is then from 40 to 100 decibels, which is quite a reasonable range of numbers, and much easier to get used to than the same sounds expressed in terms of actual intensities. (The numbers given in Table I are meant to be illustrative rather than exact; different listeners will vary widely in their judgment of what constitutes *ppp*, *f*, and so on.)

Another equation based on Eq. (1) above may be obtained with which to find the difference in intensity levels of two sounds. If one sound has an intensity I_1, giving an intensity level L_1, and another has an intensity I_2 giving a level L_2, then

$$L_2 - L_1 = 10 \ \log \frac{I_2}{I_0} - 10 \ \log \frac{I_1}{I_0}, \qquad (2)$$

which from Eq. (12) in the Appendix becomes

$$L_2 - L_1 = 10 \ \log \frac{I_2}{I_1}. \qquad (3)$$

If $I_2 = 2I_1$, for example, we find by using the table in the Appendix

$$L_2 - L_1 = 10 \ \log 2 = 3.0, \qquad (4)$$

so that doubling the intensity of a sound increases its intensity level by 3 decibels. This is true whether the sound is loud or soft. In the same way, we find that increasing the intensity of a sound by a factor of 10 raises its level 10 decibels; by a factor of 100, 20 decibels, and so on. Working in the other direction, we find that the difference in level of 60 decibels between *ppp* and *fff* in Table I corresponds to a ratio of 10^6, or one million, in sound intensities.

Sound Pressure Level

As defined in Eq. (1), intensity level is a purely physical quantity and can be measured by physical equipment without any reference to the behavior of the ear. In practice, direct measurement of the intensity of a sound wave is difficult. It is much easier to obtain the pressure amplitude of the sound; this is what most microphones measure. The intensity of a sound wave progressing in one direction is proportional to the square of the sound pressure. Using this relationship, we may obtain from Eq. (1) an expression defining what is called the *sound pressure level*. Denoting it by L_p, we have

$$L_p = 20 \ \log \frac{P}{P_0}, \qquad (5)$$

where P is the sound pressure and P_0 is the sound pressure corresponding to the threshold intensity and has the value 2×10^{-5} newtons per square meter.

For a single sound wave, moving in one direction, the intensity level and the sound pressure level are the same. In more complicated situations, however, such as in an auditorium where sound is reflected from the walls, the two levels will not necessarily be the same. For our purposes we need not be too concerned about the difference; if a given sound pressure level is measured as, say, 90 decibels, the intensity level will not be too far from this figure.

Instruments are available commercially that measure sound pressure levels directly: both small portable instruments to be held in the hand, and larger or more elaborate laboratory models. In any of them, the electrical output of a microphone picking up the sound is put through an electrical circuit that amplifies and processes the microphone signal and puts it into an indicating meter. The sound pressure level is then given directly by the reading on the meter. The electrical circuit may be further modified so that its response to different frequencies matches somewhat the response of the ear; an instrument modified in this way is called a *sound level* meter (without the word *pressure*).

With an instrument that measures sound pressure levels, measurements may be made of the levels by musical instruments produced under somewhat standard conditions: that is, at specified distances in surroundings where there are no reflections. Some recent measurements give average intensity levels over the playing ranges of many instruments.[6] The strings and woodwinds, for example, give about 60 decibels at a distance of 10 meters when played fortissimo; the brass give about 15 decibels more. An intensity level of 60 decibels is an intensity of 10^{-6} watts per square meter; by using the inverse square law calculation illustrated in Ch. 3, we find this would be produced by a sound source power of 0.0013 watts, or 1.3 milliwatts. This value is somewhat smaller than the power outputs for musical instruments given in Table II of Ch. 3, but remember that the figures in that table are for the greatest power output for any note, whereas this value found above is the average over the playing range of the instrument. Furthermore, power outputs of instruments vary over a considerable range, so figures should be taken as representative rather than exact.

The decibel will be a convenient unit with which to describe further the behavior of the ear as a receptor of sounds. A quantity of interest to psychologists is the change in sound pressure level that is just noticeable to the listener (sometimes abbreviated *jnd*). Observations made with a number of subjects show that this quantity varies both with sound pressure level and with frequency. Fig. 2 shows curves of the just noticeable difference plotted against sound pressure level for three frequencies. It is just 1.0 decibels at the level of 30 decibels at 1000 cycles per second and drops to between one-third and one-half decibel for loud sounds.[7] The decibel is thus a unit which corresponds roughly to the amount a sound must be raised in level to be just heard as louder.

Loudness Level: Phons

Since the just noticeable difference given in Fig. 2 varies with frequency, we might expect other aspects of sound reception to do likewise; this is why we started with 1000 cycles per second as a standard frequency. For example, the hearing threshold depends on frequency.

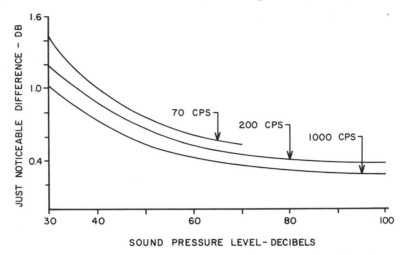

F<small>IG</small>. 2. Just noticeable difference in sound pressure level for three frequencies.

Measurements made on young people with good hearing show that the average threshold varies as given by the dotted curve in Fig. 3. At 1000 cycles per second, the actual threshold is a little above the somewhat arbitrarily chosen value of 2×10^{-5} newtons per square meter. At lower frequencies the ear is less sensitive; at 50 cycles per second, for example, the sound pressure level must be 40 decibels in order to be heard at all. Our ears undoubtedly evolved this low-frequency insensitivity so as not to be bothered by internal sounds produced by heartbeats, blood moving through veins, and so on. Above 1000 cycles per second the threshold dips down, showing that the sensitivity of the ear increases in this region, with maximum between 3000 and 4000 cycles per second. Part of this increase in sensitivity is due to the fact that the auditory canal terminating in the eardrum is a closed tube having resonance frequency in this region.

The next step would be to compare tones of different frequencies at higher levels. For example, a standard tone of level 20 decibels at 1000 cycles per second could be compared with a tone of different frequency, and the level of the latter tone adjusted until it gave the same subjective loudness sensation as the standard. In this way we could plot the lowest solid curve marked 20 in Fig. 3, which is a curve of constant *loudness level*. From this curve we see, for example, that a tone of 50 cycles per second must have a sound pressure level of about 52 decibels to give the same sensation of loudness as a 20-decibel tone of 1000 cycles per second. In the same way we may obtain other constant loudness level curves, starting with 40 decibels, 60 decibels, and so on, up to 120 decibels. These are plotted in Fig. 3.[8]

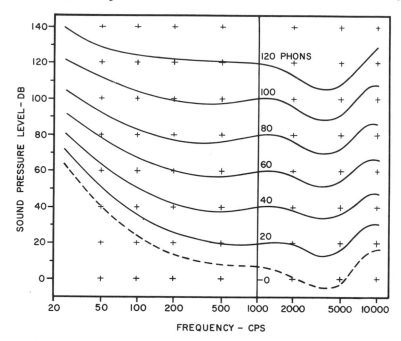

Fɪɢ. 3. Equal loudness curves relating loudness level in phons to sound pressure level in decibels.

The unit of loudness level has been named the *phon*. To connect loudness levels with sound pressure levels, the two are defined as the same at 1000 cycles per second. At any other frequency the loudness level in phons of a sound of given pressure level can be obtained from the curves of Fig. 3. For example, at 100 cycles per second a sound of pressure level 50 decibels has a loudness level of 40 phons. (It should be mentioned that the name *phon* is not used consistently; some authors use *decibel* as a unit of both intensity level and loudness level.)

The highest curve in Fig. 3, 120 phons, represents more or less the upper limit of hearing in the sense that sounds louder than this produce physiological pain in the ear, as well as possible damage to the mechanism of the ear.

Next let us see how the ear behaves when it receives two sounds simultaneously. To begin with, we shall assume the sounds to be of the same frequency, and start again with 1000 cycles per second. Suppose we add together two sounds each of intensity level 70 decibels at this frequency. (We shall ignore any difference between intensity level and sound pressure level.) Obviously we cannot add levels; this would give 140 decibels for the sum, which is absurd. What we can add are the intensities themselves, as given in watts per square meter. Two sounds of the same level will add to give twice the intensity of one, so the intensity level of the sum will be 73 decibels, as already worked out in Eq. (4).

Now let us go a step further and add two sounds of intensity levels 70 decibels and 80 decibels (still at 1000 cycles per second). The louder sound has an intensity ten times the softer one, as explained above. The sum then has eleven times this intensity, so its level L_s is

$$L_s = 10 \ \log \left(11 \frac{I_s}{I_o}\right), \tag{6}$$

where I_s is the intensity of the softer sound. Eq. (12) in the Appendix allows this to be written

$$L_s = 10 \ \log \left(\frac{I_s}{I_o}\right) + 10 \ \log \ 11. \tag{7}$$

The first term on the right is just the level 70 decibels of the softer sound, so

$$L_s = 70 + 10 \ \log \ 11$$
$$= 70 + 10.4$$
$$= 80.4.$$

We see, then, that adding a sound of level 70 decibels to one of 80 decibels gives a sum only 0.4 decibels higher, which is barely perceptible.

Two sounds of the same frequency but different from 1000 cycles per second can be combined in essentially the same way. Two sounds of loudness level 70 phons will give a sum which has a loudness level of 73 phons, since again the sum has an intensity double that of either individual sound. If we want to know the result of adding a sound of, say, 70 phons to one of 80 phons at some other frequency than 1000 cycles per second, we use the curves of Fig. 3 first to find the two intensities themselves, and find the intensity level of the sum; then we use Fig. 3 again to find the loudness level of this sum.

Loudness: Sones

Now let us go another step further and add sounds of different frequencies; for example, suppose we add a sound of 70 phons loudness level at 1000 cycles per second to one of 70 phons loudness level at 100 cycles per second. Things are immediately much more complicated. At first it might appear that we could use the curves of Fig. 3 to find the intensity of the sum, as in the preceding paragraph, and then work back again. However, although this process would give us the right intensity level of the sum, we could not work out its loudness level. This is because the curves of Fig. 3 were obtained by comparing tones of a single frequency with the standard 1000 cycles per second tone; they will not work with mixtures of tones.

In order to handle this situation, the concept of *loudness* has been developed. (Leaving off the word *level* gives the term quite a different meaning.) By asking subjects to compare sounds and decide when one sound is "half as loud" or "twice as loud" as another, we can construct a subjective scale of loudness.[9] The unit of loudness in this scale is the *sone*. According to this scheme, two sones sound twice as loud as one, 20 sones twice as loud as 10, and so on. Furthermore, the loudness in sones of individual sounds may be added to find the loudness of the sum of the sounds. As a starting point, 40 phons is arbitrarily set equal to one sone. The results of a lot of measurements on a lot of people are embodied in Fig. 4, which relates loudness in sones to loudness level in phons. (This curve should not be taken too literally in any individual case; it is an average over many subjects, and there is not only considerable variation among individuals but even the same person will make different loudness judgments at different times.)[10]

From Fig. 4 we find that increasing the loudness level by 10 phons doubles the loudness; for example, 70 phons corresponds to 8 sones,

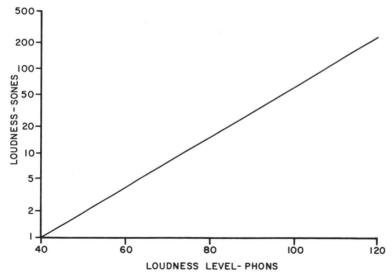

Fig. 4. Relation between loudness in sones and loudness level in phons.

and 80 phons to 16 sones. Hence if we want to know how many violins (all playing the same note at the same loudness level) will sound subjectively twice as loud as 1 violin, the answer is 10, since an increase in level of 10 phons corresponds to 10 times the sound intensity.

We may now answer the question posed above: adding 70 phons at

100 cycles per second to 70 phons at 1000 cycles per second gives a
loudness of 8 sones + 8 sones = 16 sones, which is a loudness level of
80 phons. At first this seems like a strange result: adding two sounds
of the same loudness level gives an increase in level of 3 phons if the
sounds are of the same frequency and 10 phons if they are not. The
reason for this result appears to be in the behavior of the basilar membrane. As mentioned above, a sound of a particular frequency produces
a wave traveling along the membrane; this wave has a maximum amplitude at some point whose position depends on the frequency of the
sound. The disturbance produced at this point is not concentrated at it,
but spreads over a certain length of the basilar membrane. This length
then corresponds to a range of frequencies above and below the frequency of the sound itself. This range or band of frequencies is called
the *critical band.*

Once the critical band is excited by a sound of a particular frequency, a considerable increase in sound intensity is required to produce a subjective increase in loudness; twice the intensity raises the
loudness level only three phons, as we found above. However, exciting
the basilar membrane at one point apparently does not decrease its
sensitivity at other points outside the critical band. Hence sounds of the
same loudness level—but of two different frequencies far enough apart
so as to excite the basilar membrane at two different locations—will
give a subjective impression of greater loudness than if their frequencies are the same.

We can check this idea in another way. If a complex sound is formed
of a mixture of several individual sounds of frequencies quite close together, we perceive a certain subjective loudness. If the individual
sounds are now spread apart in frequency, the loudness remains the
same until the frequency spacing exceeds the width of the critical
band; the loudness then increases.[11] Also, the individual frequencies
cannot be heard separately by the ear unless their frequency separation
exceeds the width of the critical band.[12]

Masking

We have all had the experience of being in an environment so noisy
that we are unable to hear words spoken by someone next to us. In
general, a sound that is quite audible by itself can become completely
inaudible if another louder sound is present. This situation is called
masking. In one of our calculations, above, we found that adding two
sounds whose intensities were in the ratio 10:1 produced a sound
whose intensity level was only 0.4 decibels higher than that of the
louder sound itself; this is an example of masking.

Measurements of masking are made in terms of how much the
threshold level of a given sound (the level at which it can just be

heard) is raised by the presence of a louder sound of specified level above its threshold. The latter is the masking sound, the former the masked sound. Experiments with single-frequency sounds give curves like that of Fig. 5, which shows the masking produced by a sound of

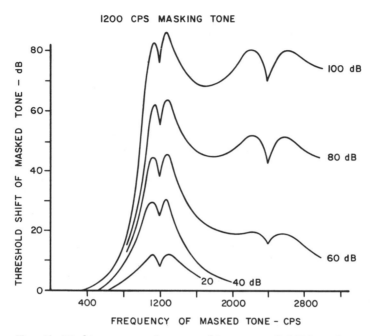

FIG. 5. Masking curves for a masking tone of 1200 cycles per second.

1200 cycles per second.[13] Low-level sounds do not mask each other unless their frequencies are fairly close together; this is shown by the curves marked 20 decibels and 40 decibels in Fig. 5. With a 40-decibel masking sound, for example, a sound of about the same frequency (of 1200 cycles per second) must be raised some 26 decibels above its normal threshold to be heard—that is, to within 14 decibels of the level of the masking sound.

When the two sounds are quite close together in frequency, beats are produced that make the masked tones easier to detect. This is the reason for the small dip in the masking curves at 1200 cycles per second. The relatively narrow frequency range in which masking occurs is again due to the fact that sounds of widely different frequencies excite separate portions of the basilar membrane, as was discussed above for the loudness of two sounds.

For loud masking tones, the situation is somewhat different. A loud sound will not mask those below it in frequency, but it will mask those of higher frequency. This is shown in Fig. 5 by the curves for masking sounds of levels 80 decibels and 100 decibels. A discussion of this effect must wait until we have discussed complex sounds in Ch. 6.

The subject of masking is of considerable importance to music. The masking of the sound of one instrument by the sound of another is of direct application to the subject of orchestration.[14] For example, when the brass section is playing fortissimo, the bassoonist cannot even hear his own instrument. Thus there is no reason for him to play at all, unless the composer does not want him to feel left out or the orchestra manager wants to get his money's worth. Masking as it affects music has not received the attention and study it should have.

Hearing Loss

Not only do our ears hear sounds; they can also be affected by sounds. We have all had the experience of being deafened momentarily by a very loud sound such as an explosion; this effect is called a *temporary threshold shift*, since it usually disappears in a short time. However, prolonged exposure to loud sounds can produce a permanent increase in the threshold of hearing that constitutes an actual hearing loss.

To check a person's hearing, a device called an *audiometer* is used. This instrument produces sounds in a pair of earphones worn by the subject. The operator can change the level of these sounds until they become just inaudible to the subject. The threshold so determined is plotted on a chart in terms of the difference in decibels between that for the subject and that for a person with normal hearing. This is done at a number of frequencies, and the results plotted on a chart called an *audiogram*.

The chart commonly used to plot audiograms is shown in Fig. 6. The frequencies are marked along the top, and the difference of the subject's threshold from the average is referred to the decibel scale at the left. The zero line at the top of the chart represents normal hearing; however, deviations above or below this line within the limits marked "normal hearing" are of no particular significance. A definite hearing loss will be shown by a pronounced dip below the zero. Thus a person who needed 100,000 times the normal intensity to hear a sound of 4000 cycles per second would have a hearing loss of 50 decibels at this frequency. We would then see a dip in the audiogram, as shown in Fig. 6 by the dotted curve.

Hearing loss may be caused in a number of ways. Continued exposure to loud sounds has already been mentioned; the recent practice of performing certain forms of popular music at high sound levels runs the serious risk of producing a permanent hearing loss in the listeners.[15] For environments in which high sound levels exist, criteria are avail-

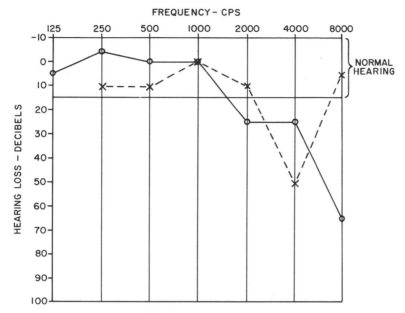

FIG. 6. Audiograms. Solid curve: presbycusis. Dotted curve: hearing
loss produced by exposure to gunfire.

able to determine whether there is any likelihood of hearing impair-
ment in the occupants.[16] Anyone who spends much time in such a noisy
environment should have his hearing tested periodically; so should any
musician.

Disease may cause hearing loss; sometimes such loss is partially
curable by surgery. We are all subject to a progressive hearing loss at
high frequencies due to the natural aging process; this is called *presbycusis*. It is illustrated in Fig. 6 as a drop in the audiogram at high
frequencies, shown by the solid curve. It takes place in a frequency
region which, as we shall see, is not important to music.

CHAPTER SIX

TONE QUALITY

The human ear is continually bombarded by sounds of all kinds. Their variety is indicated by the extensive vocabulary available to describe them—whistles, squeaks, honks, howls, and a great many others. From this infinite variety there is one kind of sound to which we shall devote most of our attention. It is the kind produced by most of our musical instruments, and is called a *tone*. This word has many other meanings in the language, and we shall use them also; but for the present it will mean a sound which lasts long enough and is steady enough for the ear to ascribe to it certain characteristics. These are loudness, quality, and pitch.

The loudness of tones was covered in the previous chapter and pitch will be covered in the next. We are now concerned with what is called the *quality* of a tone—sometimes referred to as its *timbre*—the characteristic of a tone that can distinguish it from others of the same frequency and loudness.

Partials and Harmonics

In our discussion of loudness, we dealt for the most part with sustained sounds of a single frequency, such as will be radiated by a tuning fork vibrating in its usual manner. A sound of this kind is called a *pure tone* or *simple tone*. Such tones are very important in acoustical investigations; obviously we cannot expect to understand how systems react to complex mixtures of sounds until we know how they react to single frequencies.

However, in our everyday experience, pure tones are very seldom heard, even in music. With the exception of the tuning fork, most sound sources, including musical instruments, produce complex tones that are mixtures of simple tones of various amplitudes and frequencies. The individual simple tones making up a complex tone are called *partial tones* or simply *partials*. It is the number, frequency, and amplitude of the individual partial tones that determine the quality of a given complex tone.

Most musical instruments produce steady tones that can remain unchanged in character for at least a few seconds, long enough for several hundred vibrations to take place. Wind and string players put in considerable practice to be able to produce such tones. On the other hand, some instruments produce only transient tones that die out after a certain length of time. Individual tones produced by the steady-tone instruments have a very important characteristic: they are *periodic*. Investigation of these complex tones shows that whatever the complicated vibration pattern of the sound wave, it consists of a single pattern repeated identically for a considerable number of cycles. The number of cycles produced per second is obviously the frequency of the complex tone; we shall call this the *fundamental* frequency. Fig. 1

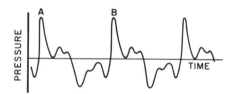

FIG. 1. An example of a vibration pattern of a complex tone.

is an example of a vibration pattern for a complex sound, being the variation in pressure during about two and a half cycles of the vibration. The pattern for one cycle, as from *A* to *B* in Fig. 1, is called the *waveform* of the vibration.

Musical tones produced by human performers will show fluctuations in the vibration pattern and frequency of the tone, due to the inability of the musician to exert perfectly steady forces or pressures on the instrument. Diligent practice develops the skill that keeps these fluctuations from being too large; however, they are never completely absent, and it would be undesirable if they were, since their presence adds a certain "life" to the tone. For our present discussion we shall disregard these fluctuations and assume that the vibration pattern repeats exactly for an indefinite number of cycles, as in Fig. 1.

For such a tone, the constituent partials must be related in a very simple way; their frequencies must be integral multiples 1,2,3,4, and so on times the fundamental frequency of the vibration. If the fundamental frequency were 100 cycles per second, for example, the partials would have frequencies 100, 200, 300, and so on cycles per second. Only if this is true will the waveform be the same for consecutive cycles, since at the end of one cycle there will have passed exactly one vibration of the fundamental, two of the next partial, and so on, so that the waveform can repeat. If we should add a frequency of 150 cycles per second to the mixture, for example, only 1.5 cycles of this vibration will have passed at the end of one cycle of the fundamental vibration, so the waveform would not repeat at 100 cycles per second.

Partials related in this simple way are given a special name: they are called *harmonics*. The partial having the fundamental frequency is called either the *fundamental* or *first harmonic*. The partial having a frequency twice that of the fundamental is the *second harmonic*, and so on.

In Ch. 4 we found that strings and open air columns can vibrate in modes whose frequencies are integral multiples of a fundamental frequency. A sound produced by a string or air column will then contain harmonics produced by these modes. A closed tube can vibrate in modes whose frequencies are integral odd multiples 1,3,5, and so on of the fundamental frequency; the sound from such a tube will then contain only odd harmonics.

The term "overtone" has frequently been used in reference to complex tones, such a tone being described as consisting of a fundamental and its overtones. This introduces a certain amount of confusion; in the sound produced by the vibrating string, the first overtone is the second harmonic. Similarly, in the sound produced by the closed tube, where the second harmonic is missing, the first overtone is the third harmonic. Because of this unnecessary confusion, it is best not to use the term "overtone" at all—especially since we do not need it.[1] The term "second harmonic" will always refer to a partial whose frequency is precisely twice that of the fundamental; this partial may or may not be present in a tone, but there is no ambiguity.

Tone Structure

To see how a complex tone is built up, let us assume that we have present in the air two single tones, one having twice the frequency of the other and, say, three quarters the amplitude. The individual waves would be as shown by the two solid curves in Fig. 2(a); they can represent either the displacement of the air molecules with time or the pressure variation with time. The resultant displacement of the air molecules (or the resultant pressure) will be the sum of the displacements (or pressures) of the individual waves, and will be as shown in Fig. 2(b); its vibration pattern is obviously more complicated than

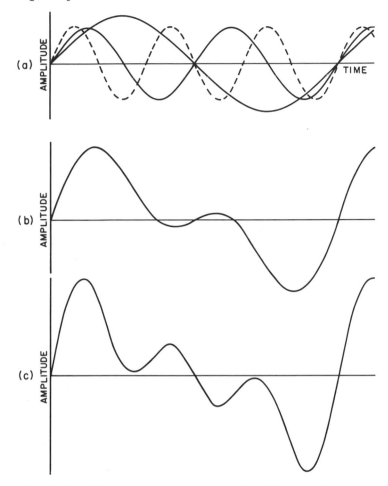

Fig. 2. Building up a complex wave. (a) The first three harmonics. (b) Sum of the first two. (c) Sum of the first three.

the sinusoidal patterns we have dealt with up to now. It is a complex sound made up of the first two harmonics shown in Fig. 2(a). Similarly, a third harmonic, as shown in Fig. 2(a) as a dashed curve, may be added to the other two. Fig. 2(c) gives the result; it has a still more complicated waveform. Adding harmonics of higher frequencies would produce a waveform of still greater complexity which, however, will still repeat at the fundamental frequency.

In Fig. 2 the individual harmonics were assumed to have their phases such that their magnitudes were zero at the same instant. This does not have to be the case; the phase of the second harmonic may have any arbitrary value at the time the fundamental is zero. We may then speak of the *relative phase* of the second harmonic with respect to the fundamental. Fig. 3 shows the fundamental and second harmonic with identical amplitudes but with different relative phases of the second harmonic. The light lines show the individual harmonics, and the heavy line the resulting complex wave. The phase of the second harmonic relative to the fundamental is zero, one-quarter cycle later and one-half cycle later in Fig. 3(a), (b), and (c) respectively. The wave-

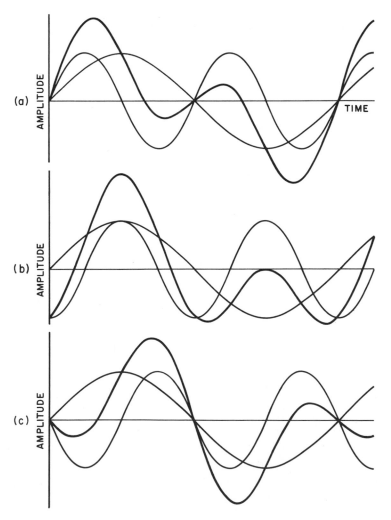

Fig. 3. Effect on the resultant waveform of changing the relative phase of the second harmonic.

form of the complex wave obviously depends on the relative phase of the second harmonic as well as its amplitude. Similarly, the waveforms of more complex waves will be quite different for different relative phases of their harmonics.

The process of adding harmonics to produce an arbitrarily complex waveform is termed *synthesis*. The converse of this is the *analysis* of a wave; any periodic waveform may be analyzed into a sum of harmonics of definite amplitude and phase, and there is only one set of such harmonics that will add up to the given waveform. With training, the ear can to some extent hear the individual harmonics of a complex sound wave.[2]

The discussion above began with sound waves, but it applies to other kinds of vibrating systems as well. In particular, complex oscillations in electrical circuits can be built up in various ways; we will discuss them in Ch. 15. One important complex wave of interest to physicists and engineers is that made up of harmonics 1,2,3,4, and so on of relative amplitudes 1, $\frac{1}{2}$, $\frac{1}{3}$, $\frac{1}{4}$, and so on, all in the same phase. It is shown in Fig. 4(a) and is called (for obvious reasons) a *sawtooth*

(a)   (b)

FIG. 4. (a) A sawtooth wave. (b) A square wave.

wave. Another is obtained by using the odd harmonics 1, 3, 5, 7, and so on with relative amplitudes 1, $\frac{1}{3}$, $\frac{1}{5}$, $\frac{1}{7}$, and so on, again all in the same phase; it is called a *square wave*. These two waves are easy to generate electrically and are important in the work that is being done on the electronic production of musical sounds.

Effect of Phase on Tone Quality

Since the detailed waveform of a complex wave depends on the relative phases of the harmonics, it might be expected that the ear would ascribe a different quality to different waveforms produced by changing the phase but not the amplitudes of the harmonics. For example, we might expect the three different waveforms of Fig. 3 to sound different. Investigation of this question over a century ago by the German physicist Ohm led him to the conclusion, called *Ohm's law* (not to be confused with another law of the same name familiar to electricians), that the perception of a complex sound depends upon its analysis into simple tones and is entirely independent of the phase relationship of these components.[3] On the basis of his researches, Helmholtz came to the same conclusion, and on the strength of his authority Ohm's Law was accepted without question for a long time. It is now firmly imbedded in most books on acoustics.

In recent years the validity of this law has been questioned, and further work done to check it. As a result, evidence has now accumulated that demonstrates conclusively that under some conditions the ear can hear changes in the phases of the harmonics of a complex tone.[4] Just what these conditions are is not yet clear; it seems to depend on how many sharp dips or peaks the waveform has.[5] In most cases, it is not a pronounced effect; the three waveforms in Fig. 3, for example, as reproduced by a loudspeaker in an ordinary room, will all sound the same.[6] For musical purposes, we may assume for now that the quality of a tone does not depend significantly on the relative phases of its harmonics; however, further investigation of this matter is needed.

Analysis of Musical Instrument Tones

Since the quality of a periodic complex tone depends mainly on the relative amplitudes of the harmonics and very little on their relative phases, we may characterize a tone by a diagram of its harmonics. In analogy to the optical case in which light is resolved into its constituent colors, the representation of the analysis of a tone into its constituent harmonics is called its *spectrum*. As an example, the spectra of the sawtooth and square waves described above are shown in Fig. 5.

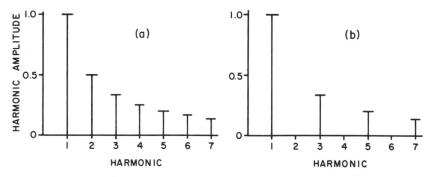

FIG. 5. Spectrum of (a) sawtooth wave; (b) square wave.

Since the tones of musical instruments have such distinctive qualities, it is natural to look for a somewhat representative sound spectrum for each instrument. One would hope that a certain spectrum would characterize the oboe, a different spectrum would characterize the horn, and so on. To this end a great deal of work has been devoted to the analysis of musical instrument tones.

The first extensive work of this kind was done with mechanical equipment.[7] The sound wave to be analyzed was made to strike a thin diaphragm. The motion of this diaphragm, caused by the varying pressure in the sound, was communicated by means of a thread to a small pivoted shaft, causing it to rotate back and forth. A small mirror was mounted on the shaft, and a beam of light was reflected from this mirror onto a moving strip of photographic film. The spot of light on the film thus moved back and forth across it as the diaphragm vibrated; the result was a plot of the pressure in the sound wave versus time, obtained as a dark trace on the film. The representation of an oscillating quantity by a graph plotted against time is frequently called an *oscillogram*. By appropriate mathematical procedures, it is possible to determine the amplitudes and phases of the harmonics in the original sound wave.

The mechanical method of analyzing sounds was cumbersome, slow, and inaccurate; much better equipment is available now for this kind of work. This equipment is electronically operated and therefore much faster; an oscillogram can be obtained in one cycle of sound and an analysis of the sound wave can be made in one second or less. In Fig. 6 we see the oscillograms of two adjacent tones sounded on a bassoon, together with their analyses into harmonic components. (The notation used to designate the tones will be explained in Ch. 8.)

The analysis of a large number of tones from musical instruments has failed to disclose any characteristic spectrum associated with a given instrument. It is found that the low tones on a clarinet are deficient in even harmonics,[8] but not all tones deficient in even harmonics sound like a clarinet. Such tones do have what might be called a "hollow" sound, and tones with many high-frequency harmonics tend to sound "brighter," but other than these vague generalizations we cannot say much. As yet it is not possible to relate a tone of given harmonic structure to a particular instrument, and sound spectra are of limited usefulness.

There are several reasons why the sound spectrum is inadequate to characterize an instrument. For one thing, the harmonic structure of a given single tone on an instrument depends on a number of factors. It changes with loudness, for example; soft tones will generally have rather few harmonics while loud tones will have many more harmonics covering a greater frequency range.[9] The harmonics in the sound radiate differently in different directions from the instrument, due to interference and diffraction effects, so the harmonic structure of a tone will depend on where it is heard. The spectrum of a tone will also depend on how the player produces it.

Furthermore, the instrument itself is not consistent throughout its range; the spectra of tones in the high ranges will be quite different

The Reception of Musical Sounds

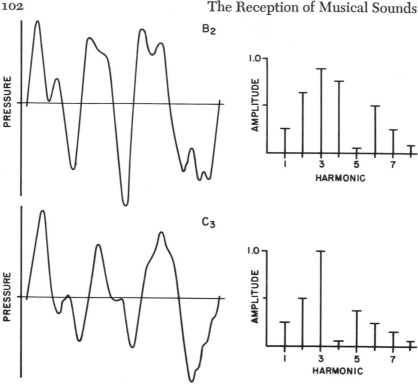

FIG. 6. Oscillograms of two adjacent tones on the bassoon, together with
their harmonic spectra.

from those in the low, and even adjacent tones in the scale of the
instrument will frequently show considerable differences in harmonic
structure. This is illustrated by the spectra in Fig. 6 for the two adjacent tones sounded on a bassoon.

In spite of the wide variation in the spectra of tones produced by a
given instrument, the ear cheerfully overlooks the differences and says
that the tones are coming from that instrument. It can do so because
there are other aspects of tone as important as the spectrum in identifying the instrument producing it. Investigations have shown, for example, that the initial transient in the instrument—the manner in which
the various partials of the tone build up to their final amplitudes—is
quite important in identifying the instrument; tones recorded without
the initial transient are much harder to identify.[10] The decay transient,
on the other hand, is quite unimportant, as we might expect. The
presence of vibrato (to be discussed below) is also helpful. Study of
tones generated synthetically by computers is furnishing more information on what factors of musical tones are important.[11] This work has

shown that the harmonic structure of a tone is quite inadequate to specify its quality.

Formants

The lack of success in correlating sound spectra with instruments has led to another theory attempting to explain tone quality. It was implied in the theory of quality outlined above that an instrument has a spectrum characterized by a particular harmonic structure, which ideally would be the same for each note of the instrument. Instead, the alternative theory suggests that each instrument has a fixed frequency region or regions in which harmonics are prominent, regardless of the frequency of the fundamental. A region of this kind is called a *formant*.

To illustrate this, Fig. 7 shows a hypothetical tone with a formant in the region 800 to 1000 cycles per second. For a fundamental of 100 cycles per second the 8th, 9th, and 10th harmonics are accentuated, as in Fig. 7(a). If the fundamental frequency is doubled, the 4th and 5th harmonics are emphasized, as in Fig. 7(b).

According to this theory, the number and positions of the formants determine the tone quality of an instrument. This is true for the human

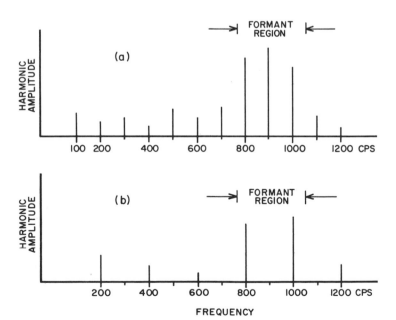

FIG. 7. Example of hypothetical tone produced by an instrument having a formant in the region 800–1000 cycles per second. (a) Fundamental of 100 cycles per second. (b) Fundamental of 200 cycles per second.

voice, as we will see in Ch. 11. For musical instruments, the situation is not clear. One investigator finds no evidence of formants in the oboe, clarinet, and French horn.[12] The bassoon, on the other hand, apparently has a fairly strong formant in the region 450–500 cycles per second; whether this is an essential part of what we think of as "bassoon tone" has not been demonstrated.[13] There seems to be no agreement as to whether or not formants exist in most musical instruments.[14]

Vibrato

In actual performance, many musical instruments do not produce tones that are as steady as the musician can make them; instead, a periodic variation of the tone—called *vibrato*—is deliberately introduced to give the tone added musical interest.

The tone vibrato is a periodic variation of the frequency of the tone from its average value. The technical term for such a periodic variation is *frequency modulation*. The frequency of the vibrato (as distinguished from the average frequency of the tone) is in the range of five to eight oscillations per second; this is the modulation frequency. The amount of variation of the frequency of the tone may be as much as three percent above and below the average value; as we will see in Ch. 8, this is a total range of a semitone.

The vibrato is usually accompanied by a variation in the amplitude of the tone vibration as well as its frequency. This is termed *amplitude modulation*. It is not an essential part of the vibrato.

Nonperiodic Tones: Noise

The above discussion of complex tones and their frequencies was based on the assumption that the tones were made up of periodic vibrations. Not all instruments produce tones of this character. For tones whose waveform does not repeat each cycle, the partials need not have frequencies that are integral multiples of the fundamental frequency, and so are called *inharmonic*. We will discuss these later in connection with the piano and percussion instruments.

If we start with a musical tone of arbitrary harmonic structure and add to it inharmonic partials to fill in the gaps between harmonics, the tone will lose its musical quality. When we have obtained a mixture of a very large number of frequencies, the sound becomes what is termed *noise*. A mixture of sounds of all frequencies that contains equal amounts of sound power in equal frequency bands of the spectrum is called *white noise*, in analogy to white light, which is a mixture of all colors. Noise of this kind has a sound resembling that produced by air escaping from a punctured automobile tire.

The musical instruments that produce periodic tones usually generate some noise of this kind also. The flute is an example; the air stream from the player's lips striking the mouth hole produces a "rushing" sound. The tones of these instruments are then not exactly periodic, and contain some inharmonic partials produced by the noise. However, the harmonic partials predominate, and the ear generally disregards the inharmonic ones.

There are other more transient sounds associated with the playing of an instrument which are included in the general term *noise*. These include the thump of piano keys, the click of wind-instrument keys, and so on. They are generally disregarded also, but are a larger part of the instrument's sound than most listeners realize.[15] It appears that the noises accompanying the sound of an instrument may serve as subconscious clues to its identification.

Chorus Effect

The quality of the sound of two or more instruments playing together is different in an important respect from that of a single instrument. A tone played on a violin, for example, will have a certain harmonic structure. If the same note is played on another violin, the tone produced will in general have a somewhat different harmonic structure, but it will still be a violin tone. If the two are played together, a different situation arises. The two tones will not have exactly the same fundamental frequency except perhaps momentarily, by coincidence. As a result, there will usually be a certain number of beats per second between the two fundamentals. There will also be twice this many beats per second between the second harmonics, and so on. Because of these beats the resulting tone will have a quality different from that of a single violin. We may go further in this direction and put together ten violins, all playing the same tone. We will then have ten slightly different fundamentals beating together, ten second harmonics doing the same, and so on. The quality produced in the tone by this slight spread in frequencies is called the *chorus effect*. Because of it, the tone of ten violins playing together is essentially different from the tone of one violin amplified to ten times the power. The practice of using many instruments in the string sections of a symphony orchestra is dictated partly by the necessity of getting a good loudness balance with the other orchestral instruments, but an additional advantage is gained in the richness of the tone the chorus effect produces. The same effect occurs in the addition of stops on the pipe organ.[16]

Subjective Tones

The ear played a prominent part in our discussion of loudness. So far, we have rather neglected it in discussing tone quality, but there is no reason for assuming it to be unimportant; instead, we should adopt a suspicious (or scientific) attitude and not absolve it from responsibility in the perception of tone quality without some investigation.

The first evidence that the ear is implicated came with the discovery, some two hundred years ago, that when two fairly loud tones were sounded together, a third tone could be heard. For example, if loud tones of 300 cycles per second and 400 cycles per second are sounded simultaneously, the ear will hear a third tone of frequency 100 cycles per second. There is no tone of this frequency present in the original sound; a frequency meter will indicate only the presence of the original sounds of 300 and 400 cycles per second. Tones not actually present in the sound but heard by the ear are called *subjective tones*. In our particular example, the frequency of the subjective tone is the difference of the two frequencies in the original sound; thus it is called a *difference tone*.

In the complex tones of musical instruments, particularly in the lower ranges, the fundamental may be weak or even absent. Complex tones may also be built up out of harmonics but with the fundamental omitted. The ear generally hears such tones as having the fundamental frequency, even though there is no actual vibration of this frequency present in the sound.[17] This *missing fundamental* effect is explained on the basis of difference tones, since any two adjacent harmonics will have a difference tone of the fundamental frequency. Thus, if a tone consists of harmonics of frequencies 200, 300, 400, and so on cycles per second, the difference tone will be 100 cycles per second for any two consecutive harmonics, and is the fundamental frequency. This effect is invoked to explain the apparent presence of bass notes in the sound from cheap radios that cannot actually reproduce anything below about 250 cycles per second.

For a long time it was believed that difference tones were caused by the behavior of the mechanical structure of the ear. To see how this could happen, let us consider the simple mechanical system in Fig. 8(a), consisting of a spring distorted in the manner shown. When it is stretched, it will act like an ordinary spring. When it is compressed, however, the close turns at one end will come into contact and can no longer move; a greater force will then be required to compress the spring further. A graph of the force necessary to produce a given displacement is then of the sort we see in Fig. 8(b); it is not a straight line, as it would be for an ordinary spring. For this reason, a system of this kind for which the displacement is not directly proportional to the force is called *nonlinear*.

If we apply a sinusoidal exciting force to this system, we will get a nonsinusoidal displacement, as in Fig. 8(c). This curve will have higher harmonic components in addition to the fundamental. The system we

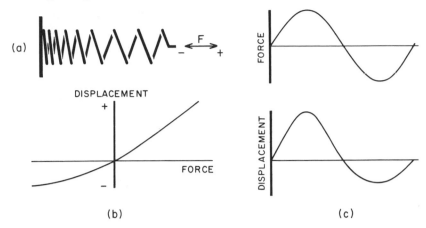

FIG. 8. (a) A distorted spring which behaves differently under compression than it does under stretching. (b) Graph of force required for a given displacement. (c) Nonsinusoidal displacement resulting from a sinusoidal force.

have used for illustration is rather simplified, but any nonlinear system will behave somewhat in this way, generating harmonics that are not present in the original excitation. The relative amplitude of the harmonics so produced will depend on the amplitude of the excitation; that is, the ratio of the amplitude of a given harmonic to the amplitude of the fundamental will generally increase as the excitation increases.

From this discussion it follows that if the mechanism of the ear is nonlinear, a pure tone will be perceived as a complex tone because of the harmonics the ear mechanism generates; these are called *aural harmonics*. Furthermore, a given aural harmonic will appear at a certain loudness level of the pure tone and increase fairly rapidly in level as that of the pure tone raised. Curves have been published giving the levels of the aural harmonics plotted against the level of the pure tone causing them.[18]

In Ch. 5 we stated that a loud tone does not mask tones lower than it in frequency, but does mask higher tones; this was shown in Fig. 5 of Ch. 5. This may be explained as masking of the high-frequency tone by the aural harmonics of the loud low-frequency masking tone. The dips in the 80- and 100-decibel curves at high frequencies may be explained as due to beats between the masked tone and the aural harmonics.

The nonlinearity of the ear will have further consequences. It can be shown that if two pure tones of frequencies f_1 and f_2 (with f_2 the higher) are heard simultaneously, there will be produced not only the aural harmonics of f_1 and f_2, but other frequencies as well; one of these will be the difference frequency f_2-f_1, so the difference tone is explained on this basis. Another one that will be produced is f_2+f_1; this is called a *summation tone*. Helmholtz claims to have heard such tones, so their existence has been accepted as proved. Theoretically, there should be other difference and summation tones as well—of frequencies $2f_1-f_2$, $3f_1-2f_2$, $2f_1+f_2$, and so on. All of these are lumped together under the general heading of *combination tones*.

The explanation of subjective tones in terms of the nonlinearity of the ear is quite believable, and it was accepted for a long time. As often happens in scientific investigations, however, recent work has thrown everything into confusion. For one thing, the mechanical portion of the ear has been found to be linear, so that any nonlinearity present must be in the cochlea and the associated nerve connections.[19] Others besides Helmholtz have searched for summation tones but have not heard them; it seems generally agreed now that they do not exist. The combination tone $2f_1-f_2$ is not only heard, but is audible at levels far below where it should be on the basis on the nonlinearity explanation.[20] Some authorities in the field of hearing deny that aural harmonics exist at all.[21]

It now begins to appear that the ear, for reasons not yet well understood, can hear a periodic change in the pattern of a sound stimulus; if the repetition frequency of this change is fast enough, the ear will hear a sound of this frequency, even though such a sound is not present in the original stimulus. Thus if we listen to two tones of frequencies 500 and 501 cycles per second, we will hear the beat as a periodic change in loudness of frequency one per second. If we sound together 250 and 501 cycles per second, we again hear a periodic change in the sound with a frequency one per second; this is the frequency at which the wave pattern repeats.[22] For frequencies of 400 and 500 cycles per second, the "beat frequency" would be 100 cycles per second; this is too fast to be called a beat, but is the frequency at which the wave pattern repeats. The ear thus hears this as the difference tone. For frequencies 250 and 450 cycles per second, the wave pattern repeats at 50 cycles per second, so the ear hears the combination tone $2f_1-f_2$ mentioned above. Finally, a complex tone with harmonics of, say, 200, 300, 400, and so on cycles per second will repeat at 100 cycles per second, so the ear will supply the missing fundamental.

For some time it was thought that subjective tones played an important part in the hearing of music, since such tones would be heard in addition to those in the music and might not be in tune.[23] However, it now appears that at the usual levels at which we listen to music, subjective tones are not important.[24]

If the reader has gained the impression at this point that we still do not know very much about what constitutes tone quality, he is perfectly correct. Further research will uncover more aspects of the problem, and we may hope that it will also provide some answers. In particular, it may someday be possible to determine what it is that differentiates a "good" tone made by a musical instrument from one that is not as "good." On this subject we are quite ignorant.

CHAPTER SEVEN

FREQUENCY
AND PITCH

We have discussed the response of the ear to the amplitude and wave-form of a musical tone, so now we need to see how the ear responds to the frequency of such a tone. A few experiments with sounds produced by vibrating objects show immediately that the frequency of a sound is closely related to what is called its *pitch*—the subjective property of a sound that enables it to be compared to other sounds in terms of *high* or *low*. Low frequencies correspond to low pitches, and conversely; however, the correspondence is not exact, since pitch is determined by other factors besides frequency.

The Audible Frequency Range

In Ch. 5 we found that as we go to lower frequency sounds, the ear becomes less sensitive; the curves of Fig. 3 in Ch. 5 illustrated this. If we go to low enough frequencies, we find that a vibration in the air is not heard at all, so for our purposes it is no longer what we would call sound. This lower frequency limit for audible sounds is difficult to determine precisely. For one thing, like most hearing phenomena, it will vary considerably from person to person. Also, it is difficult to tell when a vibration stops becoming a sound and becomes a feeling instead. Furthermore, it is difficult to produce pure tones of considerable intensity at low frequencies, and if harmonics are present in the tone, the ear will think it is hearing the fundamental. However, as a rough average figure, we may take the lower frequency limit of sound to be about 15 cycles per second.

Musically, the actual value of the lower limit is immaterial. *Pitch discrimination,* the ability to distinguish two tones of nearly the same frequency as different in pitch, is very poor near the lower limit. The lowest frequency tone of the piano is 27.5 cycles per second, and even at this higher value pitch discrimination is not too good. Hence we may consider the frequency range of musically useful sounds to start at about where the piano does—at about 27 cycles per second.

Going now to high frequencies, we find there is also an upper limit to the frequencies the ear will hear as sound. This limit is even more indefinite than the lower limit. Not only does it vary widely from one person to another, but in a given individual it will change with age. Young people can hear sounds of frequencies up to 17–18 kilocycles per second or higher; in older people this will drop to 12 kilocycles per second or lower. This hearing loss at high frequencies, called presbycusis, has already been discussed in Ch. 5 and is part of the normal aging process.

As a practical upper limit to sound frequencies, we may take an average of 15 kilocycles per second. The total frequency range of audible sounds is then 15 to 15,000 cycles per second. The ear can thus handle a frequency range 1000 to 1. This is considerably smaller than the sound pressure range it can handle which, as we saw in Ch. 5, is about 1 million to 1. As with very low frequencies, pitch discrimination is very poor at high frequencies; it is essentially nonexistent above about 10 kilocycles per second.

Frequency Range of Musical Sounds

Measurements have been made on the sounds of musical instruments to determine the frequency ranges they cover.[1] It is found that most instruments produce very little sound above 10 kilocycles per second and what is produced in this region is associated with noises such as bow scrapings, clinking of keys, and so forth. For perfect reproduction of instrument sounds, particularly for the percussion instruments, the whole audible frequency range is necessary. However, the tone quality of most instruments is very little affected by cutting off all frequencies above 10 kilocycles per second. This is why normal presbycusis is of no consequence to the musician.

The useful range of fundamental frequencies of tones produced by musical instruments is considerably less than the audible frequency range. The highest tone of the piano has a frequency of 4186 cycles per second, and this seems to have evolved as a practical upper limit for fundamental frequencies. The high-frequency region above this is necessary—in sound-reproducing equipment, for example—to accommodate the harmonics of the high tones. The working range of fundamental frequencies, the range of tones produced on musical instruments, is then approximately 27 to 4200 cycles per second.

Factors Affecting Pitch

We now need to see how sounds in the above frequency range are related to pitch. The frequency of a sound is a definite physical quantity, which can be measured on physical instruments without any reference to the ear. Pitch is our subjective evaluation of the frequency of the sound; for any given frequency there will be perceived a certain pitch, but the perception may be different in different situations, so that a specific frequency will not always have the same pitch.

To begin with, a sound must last a certain length of time in order to be assigned any pitch at all. This was implied earlier in our definition of tone; a sound of very short duration is an impulse or "click" having neither pitch nor quality. The number of cycles required to ascribe a definite pitch to a tone depends on the frequency, and is such that on the average a sound must last 0.013 seconds to give a sensation of pitch.[2] Ordinary musical tones last longer than this, so for musical purposes the effect of the duration of the tone on pitch perception can be disregarded.

The pitch of a sound of given frequency depends to some extent on its intensity. If the loudness of a pure tone is increased, a change in pitch may occur; it is generally downward at low frequencies and upward at high frequencies. A number of studies have been made of this effect, but with rather little agreement among them.[3] The effect seems to vary greatly from person to person; some do not hear it at all. On the other hand, one individual reported an apparent drop of pitch of almost a whole tone (see Ch. 8) when a sound of 200 cycles per second was raised in level from 40 to 100 decibels.[4] The effect apparently exists only for pure tones; it seems generally agreed that complex tones show no change in pitch with intensity.[5] This is fortunate for musicians, who have enough intonation problems as it is.

We usually hear sounds with two ears, and these ears are not necessarily identical. In some individuals a sound of a given frequency may produce a certain pitch in one ear and a different pitch in the other; this condition is called *diplacusis*. It is apparently negligible in normal hearing but can be produced by disease or injury to the ear.

For people with normal hearing, the effects of diplacusis and loudness on the pitch of periodic tones is small. For musical purposes, therefore, it seems reasonable to disregard these effects and use the two terms frequency and pitch as essentially synonymous. This will apply to periodic sounds, that have a definite frequency and pitch. Sounds made up of inharmonic partials will not have a definite frequency and will not in general have a definable pitch.

Pitch Discrimination

In Ch. 5 we found that the ear could discriminate changes in sound intensity level of the order of one-half decibel. By applying a little arithmetic, we find that this is a 12 percent change in sound intensity. In comparison, the ear is able to hear quite small changes in pitch. At frequencies up to about 1000 cycles per second, the ear can hear changes of about 3 cycles per second.[6] This accounts for the poor pitch discrimination at the low end of the hearing range. At 30 cycles per second, for example, a 3-cycle-per-second change amounts to 10 percent. As we will see in the next chapter, this is equivalent to almost two semitones. At 1000 cycles per second, a change of 3 cycles per second is 0.3 percent, which is equivalent to 0.05 semitones. At higher frequencies the pitch discrimination stays constant at about 0.25 percent, or 0.04 semitones. The ear is thus much more accurate in judging changes of pitch than it is in judging changes of intensity. The sensitivity of the ear to small pitch changes is quite remarkable and is still the subject of considerable research.

In Ch. 5 we discussed the subjective loudness scale constructed by having subjects decide when a sound was "half as loud" or "twice as loud" as a given comparison standard. The same procedure has been utilized in constructing a subjective pitch scale, evaluating pitches as "half as high" or "twice as high" as a given standard.[7] The unit of pitch in this scale is the *mel*, the frequency 1000 cycles per second having by definition a pitch of one mel. This scale is of interest to psychologists, having some relevance to theories of hearing, but it appears to be of no significance to music. The natural unit for a musical scale is the octave, which we will discuss in the next chapter.

Absolute Pitch

The term *absolute pitch* is the ability, possessed by some people, to name the pitch of a tone (referred to the musical scale) without having to compare it to any external standard. Musicians sometimes refer to it as *perfect pitch*. This mysterious ability has been the subject of considerable investigation.

At the present time there are two opposing points of view on the nature of absolute pitch. One group claims that most people have the ability to name the pitch of a given tone to a rough degree, that this ability can be improved with training, and that individuals with absolute pitch have carried this training to a high degree.[8] The other group claims that absolute pitch is an ability of a special kind, such that individuals possessing it will name the pitch of a given tone quickly and within a fraction of a semitone, whereas those not having it will be hesitant and will likely be off by several semitones.[9] They claim, apparently correctly, that no one originally without absolute pitch has ever been able to develop it by training to the high degree that some individuals seem to have naturally. (After a year's trial with a tuning fork, carrying it around and listening at odd times, the author was unable to develop any sense of absolute pitch.)

One proposed theory suggests that to a person possessing absolute pitch, all C's have a certain "chroma," all D's a certain different "chroma," and so on, that enable them to be identified.[10] In support of this theory it is pointed out that in naming the pitch of a tone, the individual with absolute pitch will name the right note of the scale but will sometimes put it in the wrong octave. It has been suggested that the ability can be "imprinted" in a person at an early age, just as the first moving thing a baby duck sees is imprinted in its mind as "mother," regardless of what it may be; if such imprinting does not happen at the right time in the individual's development, it can never be acquired later.[11] It is true that people with this ability are much more likely to have grown up in a musical environment. It has been suggested that we are all born with the ability, just as we are born with the ability to recognize colors, but that most of us lose it through disuse. Heredity is apparently a factor in determining whether or not a person will have the ability, but the whole subject is obviously far from settled.[12]

CHAPTER EIGHT

INTERVALS, SCALES, TUNING, AND TEMPERAMENT

While we are on the subject of frequency and pitch of musical sounds, we may take the opportunity to see how we select out of the whole range of audible frequencies those privileged frequencies to be used for practical musical purposes. It is in this area that mathematicians and numerologists (the two terms are not necessarily exclusive) have been most diligent. Numerology—that branch of the occult arts dealing with the magic of numbers—has a fascination for many, and unfortunately musicians are not exempt from its influence. This is best demonstrated by the history of musical scales.

In Ch. 7, the useful frequency range to be covered by musical instruments was given as from 27 to 4200 cycles per second. In this region there are an infinite number of frequencies, just as there are an infinite number of points on a line. The ear will not distinguish frequencies that are too close together, but with the pitch discrimination of 0.04 semitone for a good ear that was quoted in the previous chapter, there are 25 distinguishable pitches in one semitone. This will allow for a large number of pitches in the musical range, many more than have been needed. Out of this large number we select certain discrete frequencies for musical purposes; the array of chosen frequencies is called a *scale*. An individual frequency in this array is called a *note* of the scale. (The term *note*, like *tone*, has other meanings. It sometimes

refers to the symbol on the musical staff specifying the pitch and duration of a tone. It sometimes refers to the key on a piano or similar
instrument. Sometimes it is used synonymously with *tone*.)

Consonance

As long as we are willing to restrict our music to a single melody, it
does not matter much how we choose the frequencies for the notes
of the scale. However, music long ago passed the single-melody stage
and as soon as we allow for the production of more than one note at
a time, we must take into account how the combination will sound.
A combination of two or more tones of different frequencies that is
generally agreed to have a pleasing sound is called a *consonance*.

The Greek scholar Pythagoras (the same person whose name is
attached to the famous theorem in geometry) is credited with the first
observation of what constitutes consonance in a combination of two
tones. He used for this purpose an instrument called a *monochord*, consisting of a single string stretched between two supports, as in Fig. 1.
A third support can be placed anywhere under the string along its
length, thus dividing it into two segments whose lengths may have any
desired ratio.

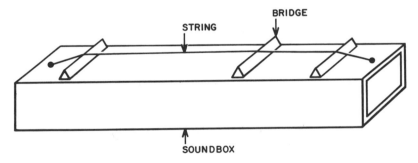

FIG. 1. The monochord.

With this instrument, it was found that the sounds from the two
segments set into vibration simultaneously were consonant when the
lengths of the segments were in the ratio 1:1, 1:2, 2:3, or 3:4. Pythagoras did not relate the string lengths to their vibration frequencies,
but we know from our discussion in Ch. 4 that the vibration frequencies of the segments on the monochord will be inversely proportional
to the string lengths. It follows that two sounds will form a consonant
combination if their frequencies are in the ratio 1:1, 2:1, 3:2, or 4:3.

Intervals

The musician will have no trouble identifying these frequency ratios.
The ratio 1:1 is rather trivial; it is two tones of the same frequency

and is called a *unison*; it would be expected to be consonant. The frequency ratio 2:1 is the *octave*, the ratio 3:2 is the *fifth*, and 4:3 is the *fourth*. This terminology refers to the keyboard on the piano. We have not as yet developed the basis for the design of the piano keyboard; however, we all know what it looks like, so it will be very useful as a convenient framework for reference purposes. (For those who find it difficult to visualize the piano keyboard without actually looking at it, a short portion of it is shown in Fig. 2.) We will discuss the frequencies actually assigned to the keyboard later.

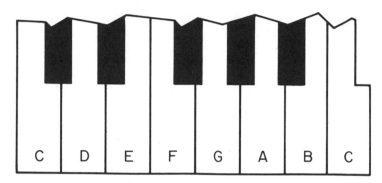

FIG. 2. A portion of the piano keyboard.

To begin, let us consider the white keys on the piano. We start with the note called C and number it 1. We then go upward (in pitch) through the successive notes D E F G A B, numbering them 2, 3, 4, and so on, until we arrive at C above as number 8. (Why we start on C rather than A is for the history of musical notation to explain.) We thus obtain the *degrees of the scale*. What is called a musical *interval* is the separation of two notes in the scale, as described by this numbering system. The interval from C, the first degree of the scale, to the C above, the eighth degree, is thus called the *octave*, although only seven steps upward are involved. Similarly, the interval from C to G, the fifth degree, is a *fifth*, although it uses only four steps upward. The interval C to F is a *fourth*. We can of course start with any other note on the keyboard and count intervals in the same way, so that not only C–G but D–A, E–B, and so on are intervals of a fifth.

From the experiments with the monochord, we identify the octave with the frequency ratio 2:1, the fifth with the ratio 3:2, and the fourth with the ratio 4:3. It is essential to see clearly the distinction between the musical meaning of an interval and its physical meaning. Musically, the term implies a difference, as the space between two points on the keyboard. Physically, an interval is a ratio of frequencies and has nothing to do with frequency differences. In the discussion of beats we used frequency differences, but they are of no significance here. As long as the frequency ratio is 2:1, the interval will be the octave. For example, two frequencies of 25 and 50 cycles per second form an octave; so do 2500 and 5000 cycles per second. The arithmetical difference between the frequencies of the first pair is much smaller than that between the frequencies of the second pair, but they are both the same interval.

The order in which we express the frequency ratios for these intervals is immaterial. The octave ratio may be expressed either as 2:1 or 1:2; to go up an octave from a given frequency, we multiply that frequency by 2, and to go down an octave, we divide it by 2, which is the same as multiplying by $\frac{1}{2}$. Similarly, the ratio for the fifth may be expressed as either 3:2 or 2:3; to go up a fifth from a given frequency, we multiply that frequency by $\frac{3}{2}$, and to go down a fifth, we multiply it by $\frac{2}{3}$.

It follows that if we add intervals on the keyboard, the frequency ratio of the sum is the product of the frequency ratios of the separate intervals. Thus if we go up a fifth from a given frequency f, we obtain the frequency $(\frac{3}{2})f$: to go up a fourth further we multiply this latter frequency by $\frac{4}{3}$, obtaining $(\frac{4}{3}) \times (\frac{3}{2})f = 2f$, so that the sum of a fourth and a fifth is an octave, as it should be. If two intervals add up to an octave, one is the *inversion* of the other. To find the frequency ratio of the inversion of an interval, we multiply the smaller figure of the interval ratio by 2. For example, taking the ratio of the fifth as 2:3 and doubling the smaller number gives 4:3 as the frequency ratio of the inversion; this is a fourth, as we would expect.

The Pythagorean Scale

With the octave, fourth, and fifth as consonant intervals expressed by the above frequency ratios, let us see how we can build up a scale of notes within the octave with the idea of obtaining as many fourths and fifths as possible. We start with the note we call C, which we will assume to have a frequency f. An octave above this will be another C, whose frequency will be $2f$. First we will add a note between these two by going down a fifth from the higher C. This is the same as going up a fourth from the lower C, so the frequency of this new note is $(\frac{4}{3})f$. Let us call this note F. Going down another fifth from F will get us on a black key; we wish to avoid these for now, so from this point on we will start on the lower C and go up by fifths. The first step will give the note which we name G and which will have a frequency $(\frac{3}{2})f$. Another fifth up from G gets us out of the octave, so we obtain the next note by instead going down a fourth from G; this is equivalent to going up a fifth and then down an octave. This gives us the note D of frequency $(\frac{3}{4}) \times (\frac{3}{2})f = (\frac{9}{8})f$. We obtain the next note A by going up a fifth from D; it has the frequency $(\frac{3}{2}) \times (\frac{9}{8})f = (\frac{27}{16})f$.

At this point we have the scale of notes together with their frequencies as shown in Fig. 3. This five-note scale, the so-called *pentatonic*

Note:	C	D	F	G	A	C
Frequency:	f	$\frac{9}{8}f$	$\frac{4}{3}f$	$\frac{3}{2}f$	$\frac{27}{16}f$	$2f$

Fig. 3. A pentatonic scale.

scale, occurs in many otherwise unrelated musical cultures, and appears to be a natural outgrowth of the existence of fourths and fifths as consonances.[1]

There are still two holes to fill in the scale. Before we do so, let us throw out the f as superfluous, since what concerns us are not the actual frequencies of the notes, but their relationships to one another in the scale. Hence the first C may be given frequency 1 and the subsequent notes expressed as fractions; once the scale is worked out, we may start at any frequency we wish by multiplying the scale fractions by that frequency.

If we now proceed by fourths and fifths as before, we will fill the two remaining spaces with notes we name E and B. The result is the scale of Fig. 4, in which the first row of fractions gives the relative frequencies of the notes in the scale. We may also consider these fractions to be the intervals between these notes and the starting point C.

Now let us find the intervals between adjacent notes in this scale. We can do this by finding the ratio of the frequency of any note to the frequency of the adjacent note below. By doing this for each note, we get the fractions shown in the second row of Fig. 4. There are two dif-

Note:	C	D	E	F	G	A	B	C
Frequency:	1	$\frac{9}{8}$	$\frac{81}{64}$	$\frac{4}{3}$	$\frac{3}{2}$	$\frac{27}{16}$	$\frac{243}{128}$	2
Interval:		$\frac{9}{8}$	$\frac{9}{8}$	$\frac{256}{243}$	$\frac{9}{8}$	$\frac{9}{8}$	$\frac{9}{8}$	$\frac{256}{243}$

Fig. 4. The Pythagorean scale.

ferent intervals—one of ratio 9:8 and the other with the rather curious ratio 256:243. A little calculation shows that the latter is closer to unity than the former (1.053 as compared to 1.125), and so represents a smaller musical interval.

The scale in Fig. 4, derived by using fourths and fifths, is called the *Pythagorean scale*. As we have derived it so far, it goes up from the starting point by first two large intervals, then one small one, then three large, and finally one small. It therefore fits the pattern of the white keys on the piano keyboard and may be termed the Pythagorean *diatonic* scale. The interval 9:8 is called the Pythagorean *whole tone*. The smaller interval 256:243 occurring in the scale is the Pythagorean *diatonic semitone*.

Since the piano has black keys inserted in the whole tone spaces between white keys, let us see what we get by adding the chromatic notes to the corresponding spaces in the Pythagorean scale. The note F#, a fourth below B, will be $(\frac{3}{4}) \times (\frac{243}{128}) = \frac{729}{512}$. (The numbers in the ratios keep getting larger.) The interval F#–G will then be $(\frac{3}{2})/(\frac{729}{512}) = \frac{256}{243}$, and so is a diatonic semitone, as we would expect. However, the interval F–F# is $(\frac{729}{512})/(\frac{4}{3}) = \frac{2187}{2048}$. This interval is the *Pythagorean chromatic semitone,* and it is a little larger than the diatonic semitone. We thus find that this scale has two different sizes of semitones. This is an awkward result, but it is a necessary consequence of the arithmetic.

This is not the only difficulty. If we continue to move in steps of a fifth up or a fourth down (to stay within the octave), we will eventually get to B♯ which is the *enharmonic equivalent* to C, and these two notes are not the same relative frequency. The easiest way to see this is to go up by whole tone steps from C to D [relative frequency $\frac{9}{8}$], to E [relative frequency $(\frac{9}{8}) \times (\frac{9}{8}) = \frac{81}{64} = (\frac{9}{8})^2$] to F# [relative frequency $(\frac{9}{8}) \times (\frac{81}{64}) = \frac{729}{512} = (\frac{9}{8})^3$] and so on, arriving after six steps at B# of relative frequency $(\frac{9}{8})^6$. The C an octave above the starting point will have the relative frequency 2. The interval B♯–C is then $(\frac{9}{8})^6:2$, which works out to be the magnificent ratio 531441: 524288, so that B♯ is higher than its enharmonic equivalent C. We find the same result if we calculate the interval between the chromatic semitone and the diatonic semitone, obtaining $(\frac{2187}{2048})/(\frac{256}{243})$, which works out to the same ratio. This interval is called the *Pythagorean comma*.

Cents

It is obvious by now that we are becoming swamped by arithmetic, and that the numbers we obtain are quite meaningless unless we can give them some musical significance. To do so, we may again make use of the scale of the present-day piano keyboard as something quite familiar and whose scale we shall presently work out. Two adjacent notes in this scale form a tempered semitone, which we will define later. However, since we are quite familiar with this semitone from our musical experience, it will be convenient to express intervals in terms of fractions of a semitone. Decimal fractions are most convenient, and we will find it most useful to use a unit comprising $\frac{1}{100}$ of a tempered semitone, called a *cent*. We may, then, express any interval in cents; in our present-day piano scale, the octave, being 12 semitones, is 1200 cents while the fifth, being 7 semitones, is 700 cents, the whole tone, 200 cents, and so on.

The number of cents in any musical interval may be easily calculated from the frequency ratio for that interval by using the method of calculation in the Appendix. For example, we stated in Ch. 7 that a good ear can hear a frequency change of about 3 cycles per second at 1000 cycles per second. Two notes of frequencies 1000 and 1003 cycles per second form an interval whose frequency ratio is 1.003. Applying the calculation to this interval, we find that it amounts to five cents. The cent itself thus represents an interval somewhat smaller than the limit of frequency discrimination in the ear, so we do not need to calculate intervals to an accuracy greater than one cent.

Now let us find the number of cents in the intervals obtained for the Pythagorean scale. The Pythagorean whole tone 9:8 works out to be 204 cents, or practically the same as our present whole tone. The Pythagorean diatonic semitone 256:243 comes out as 90 cents, a little narrow compared to the present semitone. The chromatic semitone 2187:2048, on the other hand, comes out as 114 cents, a little wider. The Pythagorean comma 531441:524288, which we found was the interval between a note and its enharmonic equivalent, works out as 24 cents, or very nearly one-quarter of a semitone. These intervals expressed as cents are much simpler and much more meaningful than in their ratio form.

Another great advantage of working with intervals expressed in cents is that we can add and subtract them, as shown in the Appendix, instead of having to multiply and divide the fractions representing the intervals. As an example of how calculations of intervals are simplified, we may find the Pythagorean comma by taking the difference between a chromatic semitone, such as E–E♯, and a diatonic semitone, such as E–F; this gives 114–90 = 24 cents, as found above.

Since B♯ is not identical to C, it follows that if we go down a fourth from E♯ and obtain F♯♯ (double sharp), this note will not be identical to G. It is easy to show that no matter how far we carry the process of moving upward by fifths or downward by fourths, we will never get exactly back to the starting point C. To see this, we note that the intervals always come out to be $(3)^m:(2)^n$, where m and n are integral numbers. The ratio can never be unity, since there is no power of three that is exactly equal to some power of two.

The Major and Minor Thirds

As long as we can satisfy our musical tastes with only the intervals of fourths and fifths, the Pythagorean scale will serve very well. However, in the evolution of music long ago there were added two new consonant intervals, which have become very important and which fit logically into the scheme of small whole numbers characterizing the intervals we have already used. These are the *major third,* with the frequency ratio 5:4, and the *minor third,* with the ratio 6:5. Associated with these will be their inversions, the *minor sixth,* with the ratio 8:5, and the *major sixth,* with the ratio 5:3. We will shortly want to distinguish intervals of the same kind in different scales, so the intervals formed by the ratios of small whole numbers will be called *just* intervals. The ratio 5:4 is then a just major third, and similarly for the others.

Now if we examine the Pythagorean scale of Fig. 4, we find that it does not contain any just thirds, major or minor. The Pythagorean major third, such as C–E, formed of two whole tones, has the ratio 81:64. Similarly, the Pythagorean minor third, such as D–F, formed of a whole tone plus a diatonic semitone, has the ratio 32:27. The Pythagorean major third is sharper than the just major third by the interval 81:80, since $(\frac{81}{64})/(\frac{5}{4}) = \frac{81}{80}$. Similarly, the Pythagorean minor third is flatter than the just minor third by the same interval, since $(\frac{6}{5})/(\frac{32}{27}) = \frac{81}{80}$. The interval $\frac{81}{80}$ is called the *syntonic comma.*

As before, it will be helpful to express these intervals in cents, using the method given in the Appendix. The syntonic comma $\frac{81}{80}$ works out as 22 cents. The Pythagorean major third $\frac{81}{64}$, being two Pythagorean whole tones, is 408 cents. The just major third 5:4 works out to be 386 cents, so that the Pythagorean major third is sharper than the just major third by the syntonic comma of 22 cents. Similarly, the Pythagorean minor third, a whole tone of 204 cents plus a diatonic semitone of 90 cents, is 294 cents. The just minor third works out to be 316 cents, so that the Pythagorean minor third is flatter than the just minor third by 22 cents.

Meantone Tuning

As the thirds came into more widespread use, musicians found that the Pythagorean major thirds sounded unpleasantly sharp and the minor thirds flat. As a result, numerous scales were proposed, formed by altering the Pythagorean scale in various ways, whose object was to make the thirds sound better. Thus we obtain the major third C–E by going up four steps of a fifth each and then dropping back two octaves. If each of these fifths is flattened by one-quarter of the syntonic comma, the resulting major third will be just, since it will be one comma flatter than the Pythagorean major third. A scale built up in this way is called *one-quarter comma meantone tuning.* In this scale the intervals C–D and D–E are both Pythagorean whole tones flattened by half a comma; this puts D halfway between C and E and is the reason for the name *meantone.* There are other varieties of meantone tunings besides the one-quarter comma tuning, but we will not bother with them.[2]

To compare this meantone scale with the Pythagorean scale, a notation using superscripts is very useful. All notes obtained by moving in steps of a just fifth (exactly 3:2) will have the same superscript, and the value of the superscript will be the fraction of a comma that the note has been raised or lowered (depending on the sign) from the value in the Pythagorean scale. On this basis, the Pythagorean scale would be written as shown in Fig. 5(a) since all the fifths are exact. The one-quarter comma meantone diatonic scale is then given by Fig. 5(b). In this scale, G is lowered by one-quarter comma, so the interval C–G is flat by that much. The note F has been raised by one-quarter

(a) C^0 D^0 E^0 F^0 G^0 A^0 B^0 C^0

(b) C^0 $D^{-\frac{1}{2}}$ E^{-1} $F^{+\frac{1}{4}}$ $G^{-\frac{1}{4}}$ $A^{-\frac{3}{4}}$ $B^{-\frac{5}{4}}$ C^0

FIG. 5. Superscript notation. (a) The Pythagorean scale.
(b) The one-quarter comma meantone scale.

comma, so the interval F–C is also flat by that much. The major thirds C–E, F–A, and G–B have all been flattened by one comma as compared to the Pythagorean major third, and so are just. The minor thirds D–F, A–C, and B–D have all been sharpened by three-fourths of a comma, so are reasonably good, though not exactly just.

To express these intervals in cents, we need first to find the value of the just fifth 3:2. This works out to be 702 cents. One-quarter of a syntonic comma is a little more than 5 cents, so the meantone fifth is 697 cents. The meantone major third is of course the same as the just major third, or 386 cents. The meantone minor third is $294 + 17 = 311$ cents, or 5 cents flatter than the just minor third.

We may add the black notes in the meantone scale by going up in flattened fifths for the sharps and down for the flats. It was customary

$C\sharp^{-\frac{7}{4}}$ $E\flat^{+\frac{3}{4}}$ $F\sharp^{-\frac{3}{2}}$ $G\sharp^{-2}$ $B\flat^{+\frac{1}{2}}$

C^0 $D^{-\frac{1}{2}}$ E^{-1} $F^{+\frac{1}{4}}$ $G^{-\frac{1}{4}}$ $A^{-\frac{3}{4}}$ $B^{-\frac{5}{4}}$ C^0

$A\flat^{+1}$

FIG. 6. A complete quarter-comma meantone scale.

to carry the scale to three sharps and two flats. The result is shown in Fig. 6.

This scale gives good intervals as long as we do not go too far away from the musical keys with three sharps and two flats. If we do, there is trouble. The enharmonic note to G♯ is the A♭ obtained by going down a meantone fifth from E♭. As seen in Fig. 6, A♭ is one comma or 22 cents sharper than the Pythagorean A♭, and G♯ is two commas or 44 cents flatter than the Pythagorean G♯. We saw above that the Pythagorean G♯ was 24 cents sharper than the A♭, so the meantone G♯ is $44 + 22 - 24 = 42$ cents flatter than the meantone A♭. This is almost half a semitone and is called the *diesis*. As a result, the interval G♯–E♭, which we ordinarily think of as a fifth on the piano, is $697 + 42 = 739$ cents, or more than a third of a semitone sharp. This interval was called the *wolf fifth* and was obviously to be avoided as much as possible. In spite of the difficulties with the wolf interval, the system of meantone tuning was serviceable enough to last for quite some time; in fact, it was used on pipe organs until about the middle of the nineteenth century.

Since G♯ and A♭ differ by the diesis in the meantone scale, it is reasonable to think of providing two separate keys for these two notes; in the organ, for example, these two keys would be connected to two separate sets of pipes. Similarly, we could divide E♭ into two keys to provide a D♯. Organs with split keys of this sort have actually been built, but the difficulties of playing on such complex keyboards have made progress in this direction impractical.[3]

The Just Scale

The use of major and minor thirds resulted in the development of a very important three-note combination. If above a just major third of ratio 4:5 we add a just minor third of ratio 5:6, we get a three-tone combination whose relative frequencies we may denote by the proportion 4:5:6. This is the *major triad,* which has been the foundation of Western music for several hundred years. Any three tones whose frequencies are in this proportion will form a major triad; for example, the proportion 200:250:300 can be reduced to 4:5:6 by dividing each number by 50. (Such a proportion is unchanged if all three numbers in it are multiplied or divided by the same number.) Similarly, the proportion 1:(5⁄4):(3⁄2) is a major triad.

F	A	C				
		C	E	G		
				G	B	D
$\dfrac{2}{3}$	$\dfrac{5}{6}$	1	$\dfrac{5}{4}$	$\dfrac{3}{2}$	$\dfrac{15}{8}$	$\dfrac{9}{4}$

Fig. 7. Building the just scale.

The major triad can be used as the basis for building up a scale. Suppose we start with the note F in our scale framework and construct on it the major triad F–A–C. We then use the top note of this triad to build the triad C–E–G, and finally on this build the triad G–B–D. This will give the note pattern in Fig. 7. If we start with the low C as frequency 1, as before, we can calculate the relative frequencies of the remaining notes by using the proportion 4:5:6 for the triad; we see the resulting numbers in Fig. 7. Now if we raise F and A by an octave, add another C an octave above the first, and lower D by an octave, we

Note:	C	D	E	F	G	A	B	C
Frequency:	1	$\frac{9}{8}$	$\frac{5}{4}$	$\frac{4}{3}$	$\frac{3}{2}$	$\frac{5}{3}$	$\frac{15}{8}$	2
Interval:		$\frac{9}{8}$	$\frac{10}{9}$	$\frac{16}{15}$	$\frac{9}{8}$	$\frac{10}{9}$	$\frac{9}{8}$	$\frac{16}{15}$

FIG. 8. The just diatonic scale.

will obtain the scale shown in Fig. 8. The relative frequencies of the notes of this scale are given by the top line of fractions in the figure. Since this scale is built up of just triads, all the fifths and major thirds in it that were parts of the component triads will be just.

If we examine the just scale further, we see that there are two other triads: E–G–B with the proportion $(\frac{5}{4}):(\frac{3}{2}):(\frac{15}{8}) = 10{:}12{:}15$, and A–C–E with the proportion $(\frac{5}{3}){:}2{:}(\frac{5}{2}) = 10{:}12{:}15$. The ratio $10{:}12 = 5{:}6$, so the lower interval is a minor third; and $12{:}15 = 4{:}5$, so the upper interval is a major third. This triad is thus the familiar *minor triad*.

Up to this point the scale looks quite good; it has three major triads and two minor triads, with all their intervals just. For this reason it is called the *just scale*. However, we do not have to look very far to find difficulties. The interval D–A, for example, has a frequency ratio of 27:40 and so is not a just fifth, but a comma flat; the interval D–F is 27:32, and so is a Pythagorean minor third instead of a just minor third. The triad D–F–A is thus out of tune.

As we look further, the difficulties increase. If we work out the intervals between adjacent notes as we did for the Pythagorean scale, we obtain the ratios shown by the bottom line of fractions in Fig. 8. (To the numerologists, these ratios are a very important feature of the just scale. They are what is called *superparticular*, meaning that the numerator of any one of them is one unit larger than the denominator. This is somehow assumed to be a desirable property of scales. We mention it here so that no one will take it seriously.) Among these ratios we see that there is what we can call a *just diatonic semitone* of ratio 16:15. This works out to be 112 cents, and so is of reasonable size. However, there are two different sizes of whole tones. One has the ratio 9:8, as in the Pythagorean scale, which is 204 cents. The other has the ratio 10:9, which is 182 cents.

Having two different sizes of whole tones is an annoying complication. However, it might be tolerated if other advantages of this scale outweighed the disadvantages of, say, the meantone scale. Unfortunately, this is not the case. Suppose we finish this scale by adding sharps and flats as in the meantone scale. We may do this in various ways. If we require that D–F♯ be a just major third, and that the minor triads on C and G and the major triads on E and A be just, we

$$\text{C}\sharp^{-2} \qquad \text{E}\flat^{+1} \qquad\qquad \text{F}\sharp^{-1} \qquad \text{G}\sharp^{-2} \qquad \text{B}\flat^{+1}$$

$$\text{C}^0 \qquad\quad \text{D}^0 \qquad\quad \text{E}^{-1}\ \text{F}^0 \qquad\quad \text{G}^0 \qquad\quad \text{A}^{-1} \qquad\quad \text{B}^{-1}\ \text{C}^0$$

$$\text{A}\flat^{+1}$$

FIG. 9. A complete just scale.

get the scale shown in Fig. 9. In this figure the superscripts indicate as before the amount in commas that the notes are raised or lowered from their values in the Pythagorean scale. With this notation, the two outside notes in any just triad must have the same superscript, being a fifth apart, and since the just major third is a comma flat as compared to the Pythagorean major third, the superscript for the middle note in the just major triad must be one unit less than the two outside superscripts, as C^0–E^{-1}–G^0 in Fig. 9. Similarly, the middle superscript in the minor triad must be one unit more, as in E^{-1}–G^0–B^{-1}. If we apply this criterion to the triad F^0–$\text{G}\sharp^{-2}$–C^0, which we ordinarily think of as a (misspelled) minor triad on the piano, we see that the middle note is three commas off; the middle note that makes this a just minor triad is $\text{A}\flat^{+1}$. Hence we find that two enharmonic notes in this scale have the same diesis error that was found in the meantone scale, so the just scale is certainly no improvement in this respect.

There are further difficulties with this scale that do not occur in the meantone scale. In the common tetrachord C–D–E–F we see from Fig. 9 that the superscripts have the pattern 0,0,−1,0. If we modulate to a new key, we will want the scale in the new key to have the same pattern. This will not be the case, however; if we start on G, for example, the pattern is 0,−1,−1,0. The A in the just scale built on C is not the right note for the scale of G; we need a new A a comma sharper. Similarly, if we start on F in the C scale, the progression is 0,0,−1,+1, so we need another B♭ a comma flatter. If we go into a remote key like

F$\sharp$, the progression is -1, -2, $+1$, -1, so that not only is the G$\sharp$ off by a comma, but the B$\flat$ is sharp by the diesis error.

Because of these difficulties, the just scale has never been of any practical use. Its theoretical attraction to individuals with numerological inclinations is extremely strong, however, so much so that it has even been called the "natural" scale, as though it had some fundamental basis in nature not possessed by other scales.[4] It appears in practically every book dealing with the acoustics of music, where it has been given an emphasis it does not deserve.

The theoretical attraction of the just scale has also resulted in the invention of keyboards of various kinds which would allow playing in many keys with just intonation. Helmholtz, for example, had a two-manual harmonium (reed organ) tuned so that the two manuals had distributed between them two just scales a comma apart.[5] This gave 24 separate notes per octave and made it possible to play with just intonation in quite a few musical keys—although not all. Helmholtz was quite emphatic in asserting the superiority of music played in just intonation; his opinions have been echoed by a number of people since, even though many of them have probably never heard anything so played.[6] Various other keyboards have been devised to allow playing in just intonation in many keys.[7] However, they have not proved practical, since musicians stubbornly cling to the present one as complicated enough.

The Tempered Scale

With the further development of music in the direction of more free modulations into more remote keys, the meantone scale became much too restrictive, so it became necessary to develop a more practical scale. The scales that we worked out above all contained twelve half-steps in the octave; these steps were generally of unequal sizes. Let us see what would happen if we made all steps exactly the same size, so that the interval C–C$\sharp$ is the same as the interval C$\sharp$–D, and so forth. We denote the value of this interval by a.

	D$\flat$	E$\flat$			G$\flat$	A$\flat$	B$\flat$	
	C$\sharp$	D$\sharp$			F$\sharp$	G$\sharp$	A$\sharp$	
C		D	E	F	G	A	B	C
	a	a^3			a^6	a^8	a^{10}	
1	a^2		a^4	a^5	a^7	a^9	a^{11}	a^{12}
	1.059	1.189			1.414	1.587	1.782	
1.000		1.222	1.260	1.335	1.498	1.682	1.888	2.000

Fig. 10. The tempered scale.

If we start with C as frequency 1, as we did above, then C♯ has the relative frequency a; D is a times this, or a^2, and so on. The relative frequencies (in terms of the number a) will then be as given in Fig. 10. To determine the interval a, we now require that the octave be exactly 2:1, as before. We then have

$$a^{12} = 2, \tag{1}$$

so a must be the twelfth root of two, or

$$a = (2)^{1/12} = 1.05946. \tag{2}$$

This interval is the *tempered semitone*, and the scale formed with it is the *tempered scale*. With the numerical value of a as found above, we may calculate the relative frequencies of the notes in the scale; they are given in Fig. 10 as the bottom rows of numbers. In the earlier scales, it was found that enharmonic notes such as C♯ and D♭ were not the same, and this was a source of trouble. In the tempered scale the enharmonic notes are identical, and this trouble disappears.

We defined the cent as $1/100$ semitone. Adding 100 cents to get one semitone is then equivalent to multiplying the interval ratio for the cent by itself 100 times. If $¢$ is the ratio for the cent, then

$$¢^{100} = (2)^{1/12}, \tag{3}$$

so

$$¢ = (2)^{1/1200}. \tag{4}$$

(This has the numerical value 1.00057779; we will not need it.)

Since the semitones are all the same size in the tempered scale, it follows that all intervals will be the same, regardless of their position in the scale. Their actual numerical values are given in Fig. 10, in the bottom rows of numbers. For example, the tempered major third may be found as the interval C–E, and has the value 1.260; by comparison, the just major third is $5/4 = 1.250$. To see how good the various tempered intervals are, we can make use of the numbers already calculated. The just fifth was found to be 702 cents. The tempered fifth is seven semitones, and so is 700 cents. The difference of two cents is negligible. Since the fifths are good, the fourths will be also; the just fourth is 498 cents, and the tempered fourth is 500 cents.

The thirds are off considerably more. The just major third was found to be 386 cents, so the tempered major third of 400 cents is 14 cents sharper. The just minor third is 316 cents, so the tempered minor third of 300 cents is 16 cents flatter. These differences are not negligible; they are responsible for the continued assertions that the tempered scale is "out of tune." However, musicians have found this scale quite practical. The ability to modulate into any key makes up for its other shortcomings, and it has been quite serviceable for some two hundred years.

Other Octave Divisions

In the history of music it has been proposed from time to time that the octave be divided into more than twelve parts, mostly in the hope that intervals in the more subdivided octave would be more nearly equal to the just intervals. To see how this would work out, let us assume the octave to be divided into M equal intervals instead of 12. Each of these would then have the ratio $2^{1/M}$. If we take m of these to build a larger interval, this will have the ratio $2^{m/M}$. If n is the number of cents (in the 12-division tempered scale) contained in this larger interval, we have

$$2^{n/1200} = 2^{m/M},\qquad(5)$$

so

$$n = 1200 \; m/M.\qquad(6)$$

With Eq. (6) we can try out various values of M to see how they work. Divisions that give good results are 19, 31, and 53 steps to the octave. For example, suppose we take five steps of the 19 division. The number of cents in this interval will be

$$n = \frac{1200 \times 5}{19} = 316 \text{ cents},$$

so this is a good just minor third.

The results of various divisions are shown in Table I. From it we see, for example, that the interval formed of 8 steps of the 31 division is 6 cents flat as compared with the just minor third. The 19 division gives reasonably good thirds and fifths, but the semitone is rather large. The intervals in the 31 division turn out to be the same as the intervals in the one-quarter comma meantone tuning. The 53 division is essentially a scale of commas, since one step in this scale amounts to about 23 cents. It follows that any interval in this scale could not deviate from a given just interval by more than half a comma. Actually, the correspondence is much better, as Table I shows; the intervals are all within one cent of the just values.

This property of the 53 division was so appealing that a harmonium was actually constructed with 53 separate notes to the octave.[8] These obviously could not be crowded into an octave of our present size, which is dictated pretty much by the size of the hand, so it was necessary to spread the keys laterally, in a direction perpendicular to the length of the keyboard. In order to reduce the amount of lateral skip-

TABLE I

OCTAVE DIVISION (M)	NUMBER OF SMALL DIVISIONS (m)	CENTS	CLOSEST JUST INTERVAL	CENTS DEVIATION FROM CLOSEST JUST INTERVAL
19	2	126	Diatonic semitone (112 cents)	+14
	3	190	Small whole tone (182 cents)	+8
	5	316	Minor third (316 cents)	0
	6	379	Major third (386 cents)	−7
	11	695	Fifth (702 cents)	−7
31	3	116	Diatonic semitone	+4
	5	194	Small whole tone	+12
	8	310	Minor third	−6
	10	387	Major third	+1
	18	697	Fifth	−5
53	5	113	Diatonic semitone	+1
	8	181	Small whole tone	−1
	9	204	Large whole tone	0
	14	317	Minor third	+1
	17	385	Major third	−1
	31	702	Fifth	0

ping that would otherwise result, some notes were duplicated so that the complete octave had a total of 84 keys. This made it possible to play in any key in just intonation if one wanted to spend the effort necessary to master the keyboard. Not enough musicians have wanted to, so this keyboard remains a curiosity.

Intonation in Performance

From time to time further suggestions are advanced regarding possible new tunings and keyboard constructions. These are all based on the assumption that it is desirable to use just intonation if at all possible. In addition to arguing the superiority of this intonation, Helmholtz declared that string players, when unfettered by the necessity of playing with a keyboard instrument, actually used just intonation, and cited tests made on the playing of a famous violinist to support this statement.[9] His authority has prevailed for some time; even today, many musicians believe that good string players performing by themselves, without the piano, play the just intervals rather than the tempered.

With modern acoustical equipment it is possible to measure the intervals actually played by violinists. This has been done, and it is found that string players, both in solo performance and in ensemble, tend toward the Pythagorean intervals rather than the just intervals.[10] For example, instead of flattening the tempered major third to make it sound more like the just interval, the violinist plays it somewhat sharper. Similarly, the tempered minor third is played somewhat flat, instead of sharpening it to bring it closer to the just interval.

The same thing occurs in choral singing.[11] It has been found that choral groups sing the major thirds sharp and the minor thirds flat, contrary to the opinions of those who claim that good choral groups sing in just intonation. Occasionally one reads a review of a choral concert in which the chorus is praised for singing in "pure" or "true" intonation; one wonders what tuning the critic had in mind, if any.

It appears that the practical musicians disregard the theorists and play what sounds best, and the centuries-old arguments as to which tuning is best and which scale is the most "natural" are a waste of time.[12] (New scales continue to be invented, however.)[13] We do need a standard of some kind to which to refer musical frequencies, and for this purpose the tempered scale is simpler than any other for present musical purposes. The fact that the frequencies in the scale can be given exact values does not mean that the musician must play precisely these frequencies; he is free to vary his playing pitch in any way he needs to fit the demands of the music. The whole problem of the actual values of the intervals musicians use in performance, and how these values depend on the musical context, the deficiencies of instruments, and so forth, still needs a good deal of study.[14]

The Standard of Pitch

Since the notes in the tempered scale are determined by the arithmetic outlined above, we need to fix the frequency of only one note in the scale in order to construct the frequencies for the remaining notes. For some centuries it has been customary to take the frequency of the note A above middle C as the standard of reference. In the early days of music, there was no recognized standard frequency for this note, each village taking its own A from the local church organ; as a result the note A in one town might be several semitones different from the note A in another town. It was not until the tuning fork was invented (by John Shore, a contemporary of Handel) that a convenient and easily used standard was available. By the middle of the eighteenth century, the frequency of A had settled down to somewhere in the range 415 to 428 cycles per second; Handel's fork, for example, was 422.5 cycles per second.

Over the years the frequency of the standard A has gone up. The reason for this rise is still not established. It appears that those instruments, such as the strings, that can tune to any pitch gain an advantage in "brightness" if they are tuned sharp with respect to those instruments, such as the woodwinds, that must be built to a fixed standard of pitch. Whatever the reason, toward the end of the nineteenth century the standard A had gone as high as 455 cycles per second in England and even up to 461 cycles per second in the United States.[15] At one period in England there were even two different standards of pitch in use, and the instruments that could be used for one pitch could not be used for the other.

A change in the standard of pitch imposes considerable difficulties on musicians and particularly on the manufacturers of musical instruments, so a fixed standard is essential. Various attempts were made to establish a standard, and finally in 1953 the International Standards Organization recommended the adoption of A-440 as the standard frequency throughout the world.

Unfortunately, the recommendations of acousticians are usually ignored, so there still exists pressure to increase the frequency of A. The Los Angeles Philharmonic Orchestra tunes to A-442, as do some moving picture studios in the area, and at least one orchestra in the United States is reported to tune to A-444. A change from 440 to 444 cycles per second amounts to about 16 cents, and is not negligible; if an instrument is constructed to play in tune at A-440, its intonation will suffer if it is shortened to play at A-444.[16] Instrument manufacturers are thus quite concerned because a change in the pitch standard necessitates redesigning the instruments—with no guarantee that their tone will not suffer in the process.

Other problems result from the rise in pitch standard. Singers, for example, should be aware that today they are singing the arias in the operas of Mozart and Beethoven about a semitone higher than the pitch for which they were written.[17] The string players have problems too; an increase of one semitone is an increase of nearly 6 percent in frequency, so to raise the pitch of a string a semitone means an increase in its tension of nearly 12 percent. To accommodate this increased tension, the violins of the old Italian masters have for the most part had to be strengthened by adding stiffer bass bars.[18] This means that the tone of these instruments is not necessarily the same now as it was when they were built.

There is no lasting benefit to be gained from raising the standard of pitch; rather, it is a never-ending process. If the strings tune sharper than the woodwinds for added "brightness" or whatever, the woodwind players will naturally play sharper to compensate. This will cause the strings to go still higher. The woodwinds will again try to follow, but there is a limit to the pitch change that they can accommodate; eventually, they will have to be built to a new pitch standard. This settles nothing, however; the whole cycle now starts over again.

More investigation is needed to determine just what the pressures are that result in the continual rise in tuning pitch. Meanwhile, it is to the ultimate advantage of musicians everywhere to keep the standard at the recommended value of A-440. Any attempt to raise this standard for any short-term gain in "brightness" or whatever else should be resisted as pernicious and contrary to musicians' best interests.

Octave Notation

With A-440 hopefully settled on as the standard frequency, we can determine the frequencies of the remaining notes in the scale. From Fig. 10 we find that the relative frequency of A is 1.682. What is called *middle C* on the piano must then have a frequency $440/1.682 = 261.63$ cycles per second. By using this together with the relative frequencies of the other notes in the scale, we can find their actual frequencies. Once we obtain these, we can find the frequencies for the notes in other octaves by multiplying and dividing by 2, 4, 8, and so on. The result will be the scale of frequencies given in Table II. In order to tune a piano properly, it is necessary to know the frequencies of the notes in the tempered scale to the accuracy given in Table II. (The numbers in this table are rounded off from longer calculated values, so some apparent discrepancies appear in the last decimal place.)

At one time there was proposed a "scientific" musical scale for which the various C's were powers of the number 2—as 4, 8, 16, 32, and so forth. In this scale, middle C had the frequency 256 cycles per second. This scale was never taken seriously by musicians, and is quite obsolete. Unfortunately, the manufacturers of scientific equipment have not yet become aware of this, and continue to make tuning forks in this obsolete scale. As a result, many people who have been exposed to high school and college elementary science courses still think middle C is 256 cycles per second.

The names C, D, E, and so on for the notes in the scale have been established for a long time, and are sufficient to specify the position of the note within the octave. Unfortunately, there is no such agreement on how to specify in which octave the note is located. Since there are seven complete octaves on the piano, this leaves room for considerable ambiguity, as is obvious from Fig. 11. Musicians designate a note by its position on the musical staff. The various C's in Fig. 11 would be designated as middle C, the C in the third space of the treble clef, and so on. This is an extremely cumbersome system and of no use to a nonmusician.

Table II

Frequencies of notes in the tempered scale

C_0	16.352		C_3	130.81		C_6	1046.5
		17.324			138.59		1108.7
D_0	18.354		D_3	146.83		D_6	1174.7
		19.445			155.56		1244.5
E_0	20.602		E_3	164.81		E_6	1318.5
F_0	21.827		F_3	174.61		F_6	1396.9
		23.125			185.00		1480.0
G_0	24.500		G_3	196.00		G_6	1568.0
		25.957			207.65		1661.2
A_0	27.500		A_3	220.00		A_6	1760.0
		29.135			233.08		1864.7
B_0	30.868		B_3	246.94		B_6	1975.5
C_1	32.703		C_4	261.63		C_7	2093.0
		34.648			277.18		2217.5
D_1	36.708		D_4	293.66		D_7	2349.3
		38.891			311.13		2489.0
E_1	41.203		E_4	329.63		E_7	2637.0
F_1	43.654		F_4	349.23		F_7	2793.8
		46.249			369.99		2960.0
G_1	48.999		G_4	392.00		G_7	3136.0
		51.913			415.30		3322.4
A_1	55.000		A_4	440.00		A_7	3520.0
		58.270			466.16		3729.3
B_1	61.735		B_4	493.88		B_7	3951.1
C_2	65.406		C_5	523.25		C_8	4186.0
		69.296			554.37		4434.9
D_2	73.416		D_5	587.33		D_8	4698.6
		77.782			622.25		4978.0
E_2	82.407		E_5	659.26		E_8	5274.0
F_2	87.307		F_5	698.46		F_8	5587.7
		92.499			739.99		5919.9
G_2	97.999		G_5	783.99		G_8	6271.9
		103.83			830.61		6644.9
A_2	110.00		A_5	880.00		A_8	7040.0
		116.54			932.33		7458.6
B_2	123.47		B_5	987.77		B_8	7902.1

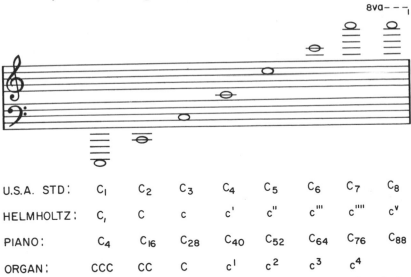

U.S.A. STD:	C_1	C_2	C_3	C_4	C_5	C_6	C_7	C_8
HELMHOLTZ:	$C_{,}$	C	c	c'	c''	c'''	c''''	c^v
PIANO:	C_4	C_{16}	C_{28}	C_{40}	C_{52}	C_{64}	C_{76}	C_{88}
ORGAN:	CCC	CC	C	c^1	c^2	c^3	c^4	

Fig. 11. The octave notation recommended by the U.S.A. Standards Association, together with some others that have been used.

Various systems have been proposed to designate octaves; some of these are shown in Fig. 11, in the rows of symbols under the staff. Helmholtz used the one in the second row of Fig. 11, and this system is still used to some extent. Unfortunately, it is a mixture of capital letters, small letters, primes, and inverted primes, and it is too complicated for musicians to remember. Piano builders have a different system, as shown in the third row of Fig. 11, in which the subscript indicates the number of the note on the piano keyboard, counting from the bottom. Organ builders have a still different system, as shown. To add to the confusion, a number of other systems have also been used at various times in the past.

The most straightforward system—and the one most easily remembered—has been proposed by the Acoustical Society of America and accepted by the U.S.A. Standards Association; it is shown in the top row of symbols in Fig. 11.[19] Since the lower limit of hearing is about 15 cycles per second but is not at all definite, let us arbitrarily start the tempered scale with a note we will name C_0 that has the frequency 16.352 cycles per second. All notes in the octave beginning with C_0 are given the subscript 0. The C an octave above this, with a frequency 32.703 cycles per second, will correspond to the first C on the piano and is called C_1. All the notes in this octave starting with C_1 are given the subscript 1. The next octave will start with C_2, and so on. This makes middle C on the piano come out C_4, with a frequency 261.63

cycles per second. The standard A is then A_4, with a frequency 440 cycles per second. The multiplicity of 4's makes this relation easy to remember. Accidentals are written $A\sharp_4$, $B\flat_4$, and so forth. The notes in the tempered scale as given in Table II are named according to this system.

With this notation, it is easy to show the playing ranges of musical instruments in terms of the notes of the scale. The range of the piano is from A_0 to C_8, between the dotted lines shown in Table II. The ranges of some of the usual orchestral instruments are shown in Table III. The upper limits depend a great deal on the skill of the player,

TABLE III

Playing ranges of musical
instruments—concert pitch

	LOWER LIMIT	APPROXIMATE UPPER LIMIT
Violin	G_3	E_7
Viola	C_3	C_6
Violoncello	C_2	E_5
Double bass	E_1	B_3
Flute	C_4	C_7
Oboe	$B\flat_3$	F_6
English horn	$E\flat_3$	$B\flat_5$
Clarinet ($B\flat$)	D_3	$B\flat_6$
Bass clarinet ($B\flat$)	D_2	F_5
Bassoon	$B\flat_1$	$E\flat_5$
Contrabassoon	$B\flat_0$	$E\flat_3$
Horn (double, F and $B\flat$)	B_1	F_5
Trumpet ($B\flat$)	E_3	$B\flat_5$
Trombone (tenor)	E_2	$B\flat_4$
Trombone (bass)	B_1	$B\flat_4$
Timpani (28″ and 25″)	F_2	F_3
Harp	$B_0(C\flat_1)$	$G\sharp_7$

so they are not very precise. The lower limits are determined by the physical dimensions of the instrument. Some of the brass instruments can go lower than the limits shown by making use of pedal tones, which we will discuss in Ch. 12; again, the usefulness of these depends partly on the skill of the player. For the strings and winds the limits are fixed, and it is silly for a composer to write, say, A_1 to be played on the bassoon, since that note is below the instrument's range.

Intonation of Instruments

The tempered scale serves as the standard for tuning pianos and organs and for building the wind and brass instruments. What the musician calls the *intonation* of one of these instruments is the degree to which its playing frequencies conform to those of the tempered scale. It depends primarily on the dimensions of the instrument as built by the manufacturer, and cannot be made perfect. It is not possible to build a wind instrument for which every note is exactly in tune; the practical necessity of using the same hole or key for more than one note requires that certain compromises be made in intonation.

An important factor in the intonation of the wind and brass instruments is the air temperature.[20] A rise in temperature has an insignificant effect on the strings. On winds and brass, however, the increased speed of sound at higher temperatures produces a considerable increase in pitch. It amounts to about 3.0 cents for each degree centigrade (1.6 cents for each degree Fahrenheit). This is partly the reason for "warming up" the instrument before playing in ensemble. It also complicates the manufacturer's problems in designing instruments, since the temperature of the air inside the instrument will not be constant, but will vary from point to point depending on how much of it is warmed by the player's breath.

Fortunately, the player of a wind or brass instrument has some control over the playing frequency, more than is generally realized. The dimensions of the instrument determine the playing frequencies fairly closely, but the player is able to "lip" the notes up or down over a small range and so compensate for the intonation deficiencies. (The acoustical reasons for this frequency control will be discussed later.) He can also alter the pitch somewhat to suit musical requirements; the pitch proper to a particular musical situation may not be exactly the frequency in the tempered scale indicated by the note in the music. This is another intangible aspect of intonation and depends on the player's musicianship.

The intonation of instruments can be checked by measuring their playing frequencies with appropriate equipment. The most useful and accurate device for this is one based on the principle of the *stroboscope*. If a rotating object is illuminated by a light which is flashed on for a short time once each revolution, the object appears to stand still. This is due to what is called *persistence of vision,* which enables us to see moving pictures as continuous motion instead of a succession of 24 separate pictures each second. The object will also appear to stand still if the light is flashed once for every 2, 3, 4, and so on revolutions of the object. If the frequency of flashing is slightly lower than the rotational frequency, the object appears to rotate slowly in the forward direction, and conversely. Objects that are vibrating rather than rotating will behave similarly.

The frequency meter under discussion utilizes this principle.[21] The sound of the musical instrument to be checked is picked up by a microphone, amplified, and made to flash a light at the frequency of the sound. This light is located behind a series of revolving translucent disks, each marked off in a pattern of black bars. There are twelve of these disks, one corresponding to each note in the tempered scale; by means of a train of gears, the disks are rotated at speeds proportional to the frequencies of the notes to which they correspond. When the frequency of the sounding instrument is exactly one of the notes of the tempered scale, a pattern of bars on the disk corresponding to that note apparently remains motionless; if the frequency is higher or lower, the pattern moves to the right or to the left. By adjusting the frequency meter to make the pattern of bars remain stationary, the deviation of the sound frequency from the tempered scale frequency may be measured. The meter reads directly in cents, and is accurate to one cent, well below the pitch discrimination of the ear.

The frequency meter is very useful for checking the intonation of instruments as manufactured or after alteration. It is also very helpful to the student as a check on his own intonation while he learns an instrument; since there exists no "natural" scale, there is no "natural" intonation. Correct intonation must be learned, and it is quite possible instead to learn incorrect intonation.

Measurements of orchestras in performance as made with the frequency meter show a fair amount of deviation on the part of individual instruments from the tempered scale values.[22] How much of this is due to the skill of the player in meeting the demands of the music, and how much to his lack of skill in playing the instrument, is still a matter for future research. The frequency meter also indicates that the oboe's A is not a very good standard of pitch, since it can easily vary as much as 20 cents above or below A-440. More reliable tuning standards than the oboe are available to orchestras willing to sacrifice tradition for accuracy in intonation.

Theories of Consonance

This chapter began with a discussion of consonant intervals. It is agreed that they exist, but why an interval with a simple integral frequency ratio such as 3:2 should be consonant is still a matter of discussion and argument.

Helmholtz proposed a theory of consonance that was based on the possibility of beats between partials of the tones forming an interval.[23] For example, if we sound together two complex tones an exact octave apart, the second, fourth, sixth, and so on harmonics of the lower tone will coincide with the first, second, third, and so on harmonics of the upper tone. If the two tones are not exactly an octave apart, there will be beats between these harmonics; for example, if the fundamentals are 100 cycles per second and 201 cycles per second, there will be one beat per second between the second harmonic of the lower tone and the fundamental of the upper; two beats per second between the second harmonic of the lower tone and the fourth harmonic of the upper, and so on. In the same manner, there will be beats between the partials of the tones forming the interval of a fifth if the frequencies are not exactly in the ratio 3:2, and similarly for other intervals. Helmholtz suggested that this beating between harmonics gives a certain roughness to the composite tone which produces dissonance when the frequencies of the individual tones depart too far from the integral relationship. It was partly on the basis of this theory that the tempered intervals, particularly the thirds, were presumed to be so much inferior to the just intervals.

It appears that the theory of beating partials is not sufficient to account for the feelings of consonance and dissonance of intervals. For example, if two tones are sounded a fourth apart and the interval then made narrower or wider by adjusting the frequency of one of the notes, the ear will accept the interval as a fourth over a range of some 20 cents sharper or flatter. Outside this range it appears to be unacceptable; if flatter than a fourth, there is a range of some 40 cents where the ear refuses to accept it as either a fourth or a major third.[24]

It is likely that a great deal of our feeling as to what constitutes consonance is a matter of musical training. One investigator, using intervals made up of pure tones and judged by nonmusicians, found that these subjects considered all intervals outside a certain minimum consonant; this minimum interval was apparently related to the critical band width in the basilar membrane of the cochlea, as discussed in Ch. 5.[25] The whole subject of consonant and dissonance, like so many in the field of musical acoustics, is quite unsettled.[26]

In the tempered scale, any given interval is exactly the same as every other interval of the same kind. It follows that, except for their height in the pitch scale, all keys will sound alike. As a consequence, the practice of ascribing certain "key colors" or certain psychological moods to different music keys has no basis in fact.[27] With string groups, there can be slight differences in the sounds of various keys because of the greater occurrence of open strings in some keys. However, for music in general, there is no acoustical or psychological reason why the key of E♭ major should sound "serious and solemn" and the key of E major should sound "expressive of joy."[28] Arbitrary emotional classifications of this kind are of no more help to music than are astrology or numerology.

PART III ⌒

THE ENVIRONMENT
OF MUSIC

CHAPTER NINE

AUDITORIUM AND ROOM ACOUSTICS

Now that we have discussed briefly the various properties of the ear and its role in the reception of musical sounds, we may think of going back to sources of sounds and study the various musical instruments. On the way, however, it will be helpful to consider in some detail how the transmission of sounds from the source to the listener is affected by the environment.

Sounds in the Open

When a source of sound is out in the open, so that the sound waves are free to spread in all directions without being obstructed, the intensity of the sound falls off as the distance from the source increases. As we saw in Ch. 3, the intensity is inversely proportional to the square of the distance from the source, so that doubling the distance reduces the intensity by a factor of four. In terms of sound intensity level, this is a reduction of six decibels for each doubling of the distance.

Situations do not occur often in which there are no obstructions in the way of the sound waves. A common situation is one in which the sound source is on the ground, which then constitutes a large obstruction; however, if the ground is a perfect reflector of sound and there are no obstacles on its surface, the inverse square law will still hold.

Even this situation is not common; usually the ground is a good absorber of sound. When we attend a band concert in a park, we usually find that the ground is covered with grass, a good absorber of sound, or it is covered with people, which are even better absorbers, as we will see. Under these conditions the sound intensity can fall off much faster than the inverse square law would indicate; the decrease in sound level may be as much as 12 decibels on doubling the distance, instead of 6 decibels.[1] This is a reduction in intensity by a factor of sixteen instead of four. The sound intensity thus diminishes rapidly as we move away from the source, and in a relatively short distance the band music is masked by the surrounding noise.

Some improvement can result from putting a band shell in back of the band. This is a reflecting surface, usually in the shape of a partial enclosure; it concentrates the sound somewhat into the area opposite the open side instead of allowing it to spread in all directions. This is equivalent to increasing the number of instruments in the band; however, it does not change the way in which the sound intensity decreases with distance. Hence to obtain satisfactory sound levels at reasonable distances in the open, it is usually necessary to use electronic amplification of some kind. This brings in new problems, which we will consider later.

Sound Enclosures

Most musical performances take place not in the open, but in auditoriums of various sizes and designs. We might think of an auditorium as a band shell extended far enough to completely enclose both the orchestra and the audience. Such an enclosure profoundly modifies the listening conditions in the audience. The subject of auditorium and room acoustics is concerned with working out the conditions under which listening will be most satisfactory.

In Ch. 3 we discussed the reflection of sound from surfaces, and saw that the intensity of a sound wave is reduced on reflection by an amount that depends on the nature of the wall. If it is perfectly rigid and smooth, no sound energy will be absorbed, so all will be reflected. Now let us assume that we are in a hypothetical room or auditorium with such perfectly reflecting walls. At some other point in the room a sound impulse is created, as by clapping the hands or exploding a toy balloon. The first sound we hear will travel to us directly from the source; it is called the *direct sound*. After a short time we hear the sound reflected from one of the interior surfaces; this will be the *first reflection*. There will follow sounds reflected from other surfaces, reflections of reflections, and so on. The situation will be somewhat as shown in Fig. 1, where the lines represent a few of the paths taken by various parts of the original wave front of the impulse. After a time the original wave will have been broken up into a great number of pieces, spread rather uniformly throughout the room and traveling in all direc-

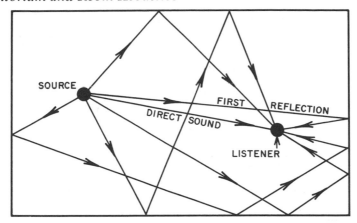

FIG. 1. Multiple reflections from the walls of a room of a single impulse produced by a sound source.

tions. What we will hear if in the room is shown in Fig. 2. First we hear the direct sound, then the first reflection, then further reflections that become smaller and closer together and finally merge into a diffuse mixture of sound waves called *reverberant sound*.

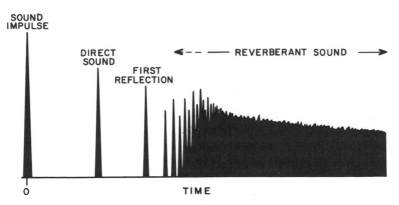

FIG. 2. Multiple reflections of a sound impulse as heard by a listener.

The intensity of the reverberant sound will obviously be of great importance to listening conditions in a room. Since we are now dealing with a mixture of sound waves traveling in all directions throughout the room, we will define the *intensity* of the reverberant sound as the power flowing into a unit area from one side only, measured as usual in watts per square meter. This intensity is what we will hear if we are in the reverberant sound. In an ideal room, the reverberant sound would be uniformly distributed and have the same intensity everywhere. In an actual room, there will be smaller or larger deviations from a uniform distribution.

In our hypothetical room with non-absorbing walls, the reverberant sound would last indefinitely. (We will have to assume for now that the listener also does not absorb sound.) The sound source has supplied a certain amount of acoustic energy to the room; since there is nowhere for this energy to go, it spreads out, fills the room, and remains there.

Reverberation Time

In actual rooms there will always be some sound absorption at the walls, so the reverberant sound will eventually disappear. By making the walls of a hard and smooth concrete, the absorption can be reduced to a minimum, and a sound can last for an appreciable fraction of a minute. Rooms built to make the sound last as long as possible are called *reverberation rooms,* and are useful in acoustic research.

To see how the sound disappears in a room with absorbing walls, let us first consider what would happen in our hypothetical room filled with reverberant sound if we were to cut a hole in one wall. Sound energy could then escape to the outside, and the reverberant sound would disappear. The rate at which energy could escape would depend on the sound intensity in the room; if the intensity were to be halved, the rate would be halved. As a result, the intensity of the reverberant sound would decrease in the manner shown by the curve of Fig. 3,

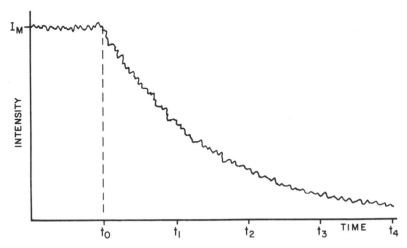

FIG. 3. Decay of reverberant sound in a room.

where the hole is assumed to have been opened at time t_0. At t_1, a certain interval of time later, the intensity has fallen to half its original value, so it is decreasing half as fast. At t_2, another equal interval later, the intensity is half its value at t_1, or one-quarter its original value. Similarly, at t_3 it will be one-eighth its original value, and so on. The curve of Fig. 3 is the same as the vibration decay curve of Fig. 3 of Ch. 2, and so may be called the *decay curve* of the reverberant sound.

Mathematically, the sound intensity will never get exactly to zero; there will always be a little left. For example, if one decides to spend every day half of what he has in the bank, his bank account will never become exactly zero; there will always be a fraction of a cent left. However, from the practical point of view, the bank will not put up with this game forever; when the bank account gets below one cent, the bank will close it out, so practically it will be zero. Similarly, there will come a point in the decay of sound in our hypothetical room when for practical purposes we can say that its intensity is zero.

The point at which the sound intensity is assumed to have reached a practical zero has been set at 10^{-6}—one-millionth—of its original intensity. In terms of sound intensity level, this amounts to a reduction of 60 decibels from the original level. The time required for the original sound to decay to 10^{-6} of its original value is called the *reverberation time*. It is a quantity of fundamental importance in auditorium acoustics.

The reverberation time of our hypothetical room with the hole in the wall will obviously depend on the size of the room and the size of the hole. If we double the volume of the room, but keep the intensity of the reverberant sound the same, there will be twice as much energy in the larger room, so if the hole is the same size, the energy will take twice as long to get out. The reverberation time T is thus proportional to the room volume V, or

$$T \propto V. \tag{1}$$

On the other hand, if we keep the room size the same but double the area of the hole, the energy will get out twice as fast, so the reverberation time is inversely proportional to the area A of the hole, or

$$T \propto \frac{1}{A}. \tag{2}$$

These may be combined into a single expression, as we have done before in Ch. 1, so we get

$$T \propto \frac{V}{A}. \tag{3}$$

Again as before, we can make this into an equation by putting in a constant of proportionality. This gives

$$T = K \frac{V}{A}.$$ (4)

To calculate the value of this constant is beyond our scope. However, the mathematicians have done it for us, and find that if we express the room volume V in cubic feet and the hole area A in square feet (instead of metric units, in deference to our architectural and engineering friends), the constant has the value 0.049. The reverberation time T in seconds is then given by

$$T = 0.049 \ \frac{V}{A}.$$ (5)

To apply Eq. (5) to an actual room is not difficult. Since all the sound in the hypothetical room that strikes the area A gets out of the room, it is essentially completely absorbed. The quantity A may then be called the total absorption in the room, measured in square feet; for convenience, we may call one square foot of opening one *absorption unit*. If instead of the opening in the wall, we had the same area of some material that absorbed all the sound striking it, the sound in the room would disappear at the same rate, and the reverberation time would be unchanged. As we saw in Ch. 3, such a material would have an absorption coefficient of unity. If now we replace this material by S square feet of some material that has an absorption coefficient a, the total absorption furnished by it would be

$$A = Sa.$$ (6)

For example, 20 square feet of material of absorption coefficient 0.6 will be equivalent to 12 square feet of opening, so is 12 absorption units.

We may extend this idea very easily; if we have S_1 square feet of material of absorption coefficient a_1, S_2 square feet of material of absorption coefficient a_2, and so on, the total absorption is

$$A = S_1a_1 + S_2a_2 + S_3a_3 + \text{---}.$$ (7)

This value can be put into Eq. (5) to work out the reverberation time. The unit of absorption is sometimes called the *sabin*, in honor of Wallace Sabine, whose pioneering work laid the foundations for our present knowledge of the subject.[2]

Optimum Reverberation Time

The presence of reverberation considerably alters the quality of both speech and music. The result is that as soon as it is created, each syllable of speech or each note of music is mixed with previous syllables or notes still present in the reverberant sound. If the reverberation time of an auditorium is too long, speech becomes scrambled to the point of unintelligibility and music so mixed up as to be muddy and unpleasant. (Playing the piano with the sustaining pedal held down produces this kind of effect.) From the standpoint of clarity in speech and music the reverberation time should be short.

However, this does not mean we should try to make the reverberation time as short as possible; we must consider another important factor. Most of our musical tones are not the impulses considered above, but steady tones that last a reasonable length of time. Suppose a musical instrument in the auditorium starts producing a tone of frequency 400 cycles per second. This amounts to producing 400 impulses per second, each of which contributes to the reverberant sound field. The intensity of the reverberant sound then builds up as shown by the upper curve in Fig. 4, where the instrument is assumed to start playing at time $t = 0$. As the intensity builds up, the rate of sound energy loss to the walls increases, and the intensity finally reaches a steady value I_M shown in Fig. 4 for which the sound power absorbed by the walls is

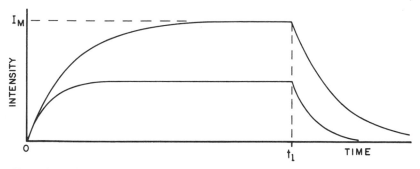

Fig. 4. Buildup and decay of sound intensity in an auditorium when a source of steady sound is present.

equal to that supplied by the musical instrument. If the instrument stops playing at time t_1, the reverberant sound decays as shown, in the way we have already discussed. The shape of the curve of intensity rise is the same as that of the decay, except for being inverted.

The maximum intensity I_M of the reverberant sound that can be produced by a musical instrument of given power will then depend on the total absorption in the auditorium. If the absorption were to be doubled, the maximum intensity would be halved, as would the reverberation time; this is shown by the lower curve in Fig. 4. In general, we can show that in an auditorium for which the reverberant sound is uniformly distributed, the maximum intensity is given by

$$I_M = \frac{P}{A'} \qquad (8)$$

where P is the sound power produced by the instrument and A' is the total absorption, but measured in units of square meters instead of square feet, so as to give the intensity in watts per square meter.[3]

We may use Eq. (8) to get an estimate of the intensity of the reverberant sound in a room. In Ch. 3 it was calculated that a sound source of one watt would produce a sound intensity of 0.009 watts per square meter at a distance of 10 feet. Suppose we want the same intensity of reverberant sound in a room with the same sound source. From Eq. (8) we find we need in the room a total absorption A' given by

$$A' = \frac{P}{I_M}, \qquad (9)$$

$$= \frac{1}{0.009},$$

$$= 110 \text{ square meters},$$

$$\approx 1200 \text{ square feet (absorption units)},$$

since one square meter is equivalent to nearly eleven square feet. If the material of the walls, floor, and ceiling of the room has an average absorption coefficient of, say, 0.3, the total area of these must be $1200/0.3 = 4000$ square feet. This would be provided by a room 40 feet by 30 feet by 11 feet high, such as a medium-size rehearsal hall.

It follows from Eq. (8) above that the shorter the reverberation time in an auditorium the greater the absorption, and the lower the sound intensity that a given source can produce. Too much absorption will then make it impossible for the musician to obtain a satisfactory sound intensity and so this condition must be avoided; the hall will be too "dead." This is strikingly shown in rooms specially constructed to have walls that absorb all the sound incident on them. Such rooms are called *anechoic*, and they are very useful for acoustical measurements. Since we are used to a certain amount of reverberation, we find that being in an anechoic room, where everything sounds muffled and quiet, has a weird effect.

The optimum reverberation time for an auditorium is thus a compromise between clarity on one hand and a satisfactory sound intensity on the other. For a given auditorium the best reverberation time will depend on the use for which it is designed as well as on its size. Fig. 5 gives a compilation of the optimum reverberation times for auditoriums

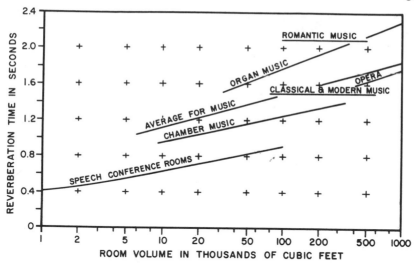

FIG. 5. Optimum reverberation time for auditoriums of various sizes and functions at a frequency of 500 cycles per second.

of various sizes and functions for a frequency of 500 cycles per second.[4] (The effect of a change of frequency will be discussed later.) It is immediately apparent from Fig. 5 that the reverberation time should be longer in larger halls; this is probably at least partly a matter of conditioning, since larger volumes will tend to produce longer reverberation times. We also see from Fig. 5 that an auditorium used primarily for music should have an optimum reverberation time longer than if used primarily for speech. This makes somewhat undesirable compromises necessary in auditoriums to be used for both speech and music. Fig. 5 indicates also that romantic music requires a longer reverberation time than classical music.

It should be mentioned that the curves of Fig. 5 are compiled from various sources, and there is by no means unanimous agreement on the numerical values given as representing the optimum values. However, we can let the authorities in this field do the arguing; for our purposes we can assume that the best reverberation time for the average fair-sized auditorium will be between 1.5 and 2 seconds.

Calculation of Reverberation Time

To calculate the reverberation time of an auditorium, we can make use of Eqs. (5) and (7). To do so, we need figures on the absorption coefficients of materials that would be used for interior wall, ceiling, and floor surfaces. Such figures are available from many sources, particularly the manufacturers of these materials. Some representative values for some common materials are given in Table I.[5] These figures are for illustrative purposes only, and should not be assumed to be precise. (There is still considerable argument about the best way to measure absorption coefficients.)[6]

TABLE I

Absorption coefficients of some building materials

	FREQUENCY—CYCLES PER SECOND					
	125	250	500	1000	2000	4000
Marble or glazed tile	.01	.01	.01	.01	.02	.02
Concrete, unpainted	.01	.01	.01	.02	.02	.03
Asphalt tile on concrete	.02	.03	.03	.03	.03	.02
Heavy carpets on concrete	.02	.06	.14	.37	.60	.65
Heavy carpets on felt	.08	.27	.39	.34	.48	.63
Plate glass	.18	.06	.04	.03	.02	.02
Plaster on lath on studs	.30	.15	.10	.05	.04	.05
Acoustical plaster, 1″	.25	.45	.78	.92	.89	.87
Plywood on studs, ¼″	.60	.30	.10	.09	.09	.09
Perforated cane fiber tile, cemented to concrete, ½″ thick	.14	.20	.76	.79	.58	.37
Perforated cane fiber tile, cemented to concrete, 1″ thick	.22	.47	.70	.77	.70	.48
Perforated cane fiber tile, 1″ thick, in metal frame supports	.48	.67	.61	.68	.75	.50

With this information, let us calculate the reverberation time of a hypothetical auditorium, which we will arbitrarily assume to be 100 ft long, 60 ft wide, and 40 ft high. The calculation will be for a frequency of 500 cycles per second. The walls and ceiling will be assumed to be plaster, with an absorption coefficient of 0.10, and the floor covered with carpet on felt, with an absorption coefficient 0.40, as given in Table I. The total absorption is then calculated from Eq. (7) as follows:

	Area, sq ft		Abs. Coeff.	Abs. Units
Floor	100 × 60	= 6000	0.40	2400
Ceiling	100 × 60	= 6000	0.10	600
Two side walls	2 × 40 × 100	= 8000	0.10	800
Two end walls	2 × 40 × 60	= 4800	0.10	480
			Total absorption	4280 ≈ 4300 units.

The volume of the room is $40 \times 60 \times 100 = 240,000$ cubic feet. The reverberation time will then be

$$T = 0.049 \times \frac{240 \times 10^3}{4300} = 2.7 \text{ sec.}$$

This would be too long for any purpose, as we can see from Fig. 5. Suppose we want to have a reverberation time of 1.5 seconds for this room; we can use Eq. (5) again to determine how much absorption to add. We find, by rearranging Eq. (5),

$$A = 0.049 \frac{V}{T}, \tag{10}$$

so that the absorption needed is

$$A = 0.049 \frac{240 \times 10^3}{1.5} = 7800 \text{ units.}$$

With 4300 units already present, we must add $7800 - 4300 = 3500$ units. From Table I we find that one-inch-thick perforated tile, for example, has an absorption coefficient of 0.70 at 500 cycles per second; 5000 square feet of this material would then supply the necessary absorption. Actually, we would need somewhat more, since the tile placed on the plaster wall will cover it, so that the plaster no longer absorbs sound. In our example, 6000 square feet of tile would supply 4200 units of absorption but would remove 600 units by covering up that much plaster, and the net gain in absorption would be 3600 units; this would be about right. The area covered would be equivalent to that of the ceiling; however, the temptation to put the absorbing material there should be resisted, for reasons that will appear later.

In the calculations above, we took into account only the absorption of the room. In an actual auditorium, however, there are other important sound absorbers: the seats and the audience. The absorption of empty seats depends on how much upholstery they have, as well as on the frequency; some representative values are given in Table II.[7] An individual member of the audience absorbs quite a bit of sound, as we also see in Table II; at 500 cycles per second, for example, each person supplies between 5 and 6 absorption units. Thus, in the above calculations, an audience of 600 to 700 people could supply the 3500 absorption units needed to bring the reverberation time down to 1.5 seconds.

Since the presence of an audience makes such a difference in absorption, it is obvious that the acoustic conditions in an empty hall during rehearsal can be completely different from those in the full hall during a performance. To reduce this difference, it is desirable to have a type of seat that when empty absorbs about as much sound per seat as

TABLE II

Sound absorption by theater seats and audience, in absorption units

| | FREQUENCY—CYCLES PER SECOND | | | | | |
	125	250	500	1000	2000	4000
Wood or metal seats, unoccupied	0.15	0.19	0.22	0.39	0.38	0.30
Cloth-covered upholstered seats, unoccupied	2.0–3.3	2.9–4.4	3.5–5.3	4.5–5.9	4.2–5.5	4.0–4.7
Audience in upholstered seats, per person	3.0–4.0	4.0–4.9	4.8–5.9	6.4–6.6	6.2–6.9	5.7–6.9

would a person sitting in it. The wood or metal seats listed in Table II would obviously be quite unsuitable. The upholstered ones listed would be much better; with seats of this kind in the auditorium, there would be little difference in the reverberation time with the hall full or empty.

Other sources of absorption sometimes need to be taken into account. The air itself absorbs sound to a certain extent, depending on its temperature and humidity.[8] The amount is negligible below 1000 cycles per second, but above this frequency it can be a factor in large auditoriums. Another source of absorption is the assemblage of pipes in a pipe organ.[9]

A word of caution is necessary at this point. The equations, graphs, and calculations given above are only approximate, and subject to considerable errors. The curves of Figs. 3 and 4, for example, apply to a room in which the sound is uniformly distributed; this can sometimes be far from the case so the curves can be considerably different from those shown. The formula for reverberation time given in Eq. (5) is only valid if the absorption is not too large, and must be replaced by another expression if there is much absorption.[10] For rooms of extreme shape Eq. (5) also does not give correct results.

Measured values of absorption coefficients may show considerable disagreement, depending on how they were measured and by whom. In particular, the figure for the absorption per person in the audience is subject to wide variations, values of from 3 units to 6.5 units being quoted by different investigators.[11] Obviously, this can make quite a difference in reverberation time calculations. One investigator has suggested that the reason for this wide variation may be because the absorption per person depends on how the audience is seated, such that a given number of people will absorb more sound if they are spread out than if they are seated close together.[12] On this point there is considerable lack of agreement. As a further complication, it appears that the absorption of an empty seat depends on whether or not it is surrounded by other seats.[13] Obviously, much still needs to be learned about sound absorption.

Measurement of Reverberation Time

The calculations of reverberation times can be quite useful in the preliminary design of an auditorium, but since they are subject to considerable error, the finished hall must be measured to see if its reverberation time is acceptable.

An estimate of the reverberation time of an auditorium can be made by simply clapping the hands and listening to determine how long the resulting sound takes to die away. For more accurate measurements, the sound of a pistol shot, or (less violently) the sound of a bursting balloon may be recorded on magnetic tape. This tape is subsequently replayed in the laboratory and the signal on it fed to a device that records the intensity level of the signal on a paper chart as a function of time. From this record the time required for the level to drop 60 decibels can be measured; this is the reverberation time.[14] A device has recently been developed that gives the reverberation time in an auditorium immediately. A pistol is fired in the auditorium, and a meter on the device reads the reverberation time directly.[15]

Instead of impulsive sounds, steady sound sources may be used. In his original work, Sabine used organ pipes, and measured with a stop watch the time required for the sound to become inaudible after the pipe was shut off. However, the use of relatively pure tones can result in complications due to resonances in the room, so it is customary to use a mixture of frequencies such as that made by a source of noise, as discussed in Ch. 6. Alternatively, a *warble tone* is sometimes used. This is a tone whose frequency is swept rapidly above and below some average value so as to produce a mixture of frequencies. In either case, the sound source is suddenly shut off and the decay of the sound in the auditorium recorded on tape and analyzed.

Since sound absorption by the audience is so important, it is very helpful to measure the reverberation time in an auditorium with the audience present. Such measurements will obviously be more difficult than in an empty auditorium. An audience may object to a pistol being fired on the stage, and it is not likely to remain quiet enough after such an outburst to allow accurate measurements of the sound decay. One method that can be used during a concert with an audience present is to record on tape the sound produced by musical chords that are percussive in character (but do not use the percussion instruments) and that end abruptly, such as certain cutoffs in Mozart's Symphony No. 40, and others.[16] It is important for this purpose that the players on those instruments that sound for some time after being played—such as harps, basses, and so on—be instructed to stop the sound quickly. A musical composition has been composed specifically for the purpose of measuring reverberation times; it contains percussive discords in which the notes are spread over various frequency ranges.[17] It can be included in a musical program without producing the audience reaction that would result from pistol shots, although there might be reactions from music critics.

Other Factors in Auditorium Acoustics

Obtaining the proper reverberation time is only the first step in achieving an auditorium which has good acoustics. It is a necessary first step, but there are a number of other factors of considerable importance. On some of these there is general agreement; on others there is still a great deal of argument.

To begin with the agreed-on factors, we note first that the reverberation time in an auditorium will depend on the frequency of the sound. In our calculation above, we used 500 cycles per second as representative. However, from Table I we see that absorption coefficients vary with frequency, generally becoming greater at higher frequencies. (There are some exceptions to this in Table I. One-quarter inch plywood on studs, for example, absorbs much more strongly at low frequencies. For this reason, the use of thin wood paneling in auditoriums should be avoided.) As a result, the reverberation time at low frequencies is generally longer than at high frequencies. This is a desirable condition, presumably because we are conditioned to it, and because it helps emphasize the bass notes in music. It appears that the increase at lower frequencies should follow a curve something like that of Fig. 6. Here T_M is the reverberation time for frequencies above 500 cycles per second. According to this curve, the value at 100 cycles per second should be about 40 percent greater. It has been found that some theatre seats have a relatively strong absorption at about 250 cycles per second.[18] Such seats could produce a dip in the curve of Fig. 6 at this frequency; this would be an undesirable condition.

Another important factor is that of uniformity of sound distribution. It is quite desirable that the sound from the auditorium stage have about the same intensity everywhere in the audience area. To accomplish this requires considerable care in the auditorium design, particularly under balconies. Any material installed for the purpose of absorbing sound should be judiciously placed. In particular, the practice of

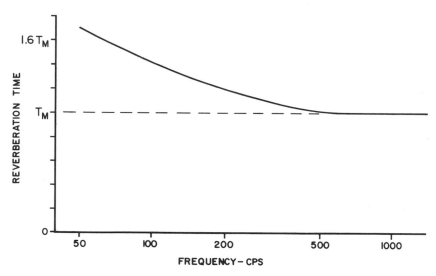

FIG. 6. Recommended variation of reverberation time with frequency.

putting absorbing material on ceilings is likely to lead to poor results, especially in smaller halls, since reflections from the ceiling are needed to carry sound to the back part of the hall.[19] Uniformity in sound distribution is affected by another factor; it has recently been found that sound from the stage traveling over the main floor seats is more strongly attenuated than might be expected simply because of the increased distance from the stage.[20] The effect is greatest at about 150 cycles per second and appears to be due to the height of the seats; it seems to be typical of most auditoriums and will result in the sound on the main floor being deficient as compared with that in the balconies. It may be alleviated by the proper slope and curvature of the floor.

Along with uniformity of sound distribution is the desirability of having enough sound to distribute. Too few instruments in too large an auditorium will result in a sound intensity too small for pleasurable listening, since it will tend to be masked by audience noise. Conversely, a large orchestra in a small hall can produce too much sound for comfort. From Eq. (8) above, the sound intensity produced in an auditorium by a source of given power depends on the total absorption; it does not depend on the volume of the hall, as one might expect. A rough rule to provide for reasonable sound intensities is to allow one instrument for each 200 absorption units.[21]

It is quite essential that the auditorium be free of echoes. In discussing reverberation, we dealt with sound reflections that in a sense were echoes, but were so closely spaced that no one stood out above the others. What is meant here is a single reflection of such intensity as to stand out clearly above the reverberant sound. Such echoes are frequently produced by flat surfaces oriented in the wrong direction. Concave surfaces are worse; very bad echoes can be produced by such surfaces, as they tend to concentrate the sound. An auditorium recently built in the Southern California area was constructed in the shape of an ellipse, with the stage at one end and with walls made of painted concrete brick, which are good sound reflectors. (This design was adopted in spite of strong objections from the acoustical consultant.) As a result, it had a very pronounced echo over most of the audience area; every loud sound on the stage was heard twice. Some (but not all) of this echo was removed by covering the concave wall with sound-absorbing material; unfortunately, the result was an increase in the total absorption to the point where the hall is now quite "dead." (It is rumored that when the acoustical consultant was asked what should be done to correct the auditorium, he answered that the best solution would be to tear it down, but this advice has not yet been followed.)

An annoying condition can occur in halls when two opposite side walls are flat, parallel, and good reflectors of sound. In this situation, a sound started in between these walls can bounce back and forth between them, producing a rapidly repeated echo known as a *flutter echo*. This condition is quite undesirable. It can be alleviated by putting sound-absorbing materials on the walls, and constitutes a further reason for not putting such material on the ceiling, since the presence of the absorbing audience insures that there will not be a flutter echo produced between floor and ceiling.

An obvious factor of importance in an auditorium is freedom from unwanted noise. It is important that a hall be so constructed that outside noises from traffic, airplanes, and so forth are excluded. However, this is matter of sound insulation, rather than sound absorption, and is therefore beyond our scope. Sources of noise inside halls need to be guarded against; air conditioners are a frequent cause of annoyance in this respect, particularly in small halls.

There are other aspects of auditorium acoustics about which there is a great deal of argument. For example, one group of auditorium acousticians claims that the time interval between the arrival of the direct sound and the arrival of the first reflection is an important factor.[22] If this interval is longer than 20 milliseconds, it is asserted that a desirable quality termed "intimacy" is lost, and the listener feels isolated from the source of sound. Unfortunately, the conditions necessary for quick arrival of the first reflection are opposed to the conditions necessary for long reverberation time; this latter requires a sufficiently large volume, and consequently puts the wall and ceiling surfaces so far away that the first reflection is slow in arriving. To get around this difficulty, some recently-built auditoriums have been provided with panels of plywood suspended from the ceiling and extending out some distance from the stage over the audience area. By putting them considerably below the main ceiling, they can then provide the necessary early reflection. However, by keeping the panels separated somewhat, sound can pass between them into the space above, so that the whole volume of the auditorium is available for reverberation.

An interesting example of an auditorium embodying this idea is the recently constructed Philharmonic Hall in New York City. This hall was provided with such suspended panels, dubbed "clouds," to provide the early reflections. Before opening the hall to the public, a week was spent in "tuning" the hall, with the help of the New York Philharmonic Orchestra.[23] For this work the presence of an audience was simulated by mats of glass fiber placed in the seats. (These were called "instant people"; not only did they absorb sound, but they were also completely quiet.) After a week of adjustments, the auditorium was judged to be acoustically good.

However, when the auditorium was opened for musical performances, complaints were heard immediately. In particular it was claimed that the hall was "dry," having too short a reverberation time, and that it was especially lacking in low-frequency reverberation. Subsequent measurements confirmed this lack, and changes were made to bring the low-frequency reverberation time up to more acceptable levels.[24]

Another group of acousticians pointed out that although the "clouds" would reflect high frequencies well, they could not be expected to reflect sounds of low frequencies and hence long wavelengths; these would pass easily through the spaces between the panels. Laboratory experiments made on arrays of panels confirmed this point.[25] In reply to this, further experiments were cited as demonstrating that the *early sound*—that arriving within 50 milliseconds of the direct sound—did not need to contain low frequencies in any amount if there were enough of them in the reverberant sound.[26] It was also claimed that the way in which the total sound energy is divided up into direct sound and reverberant sound is a very critical factor.

Other acousticians asserted that the direction from which the early sound arrives may be the important factor.[27] They suggest that if it arrives by reflection from the side walls, the listener is better able to localize the instruments in the orchestra, whereas reflections from the ceiling do not allow such localization, and a feeling of confusion results.

Acoustical Planning

It is obvious from the preceding discussion that the subject of auditorium acoustics is by no means completely understood. Research and argument are still going on, and will continue to do so; meanwhile, Philharmonic Hall has been renovated several times and will probably be subjected to more alterations in the future. However, the knowledge already available can be of great assistance in the planning of an auditorium, and an acoustical consultant should be available to an architect even before any preliminary sketches are committed to paper. Unfortunately, the practice of building the auditorium first and then calling in the acoustical consultant to correct its deficiencies is still too prevalent; this practice is about as sensible as the older one of improving the acoustics of an auditorium by stringing wires along the ceiling or piling broken wine bottles under the stage.

As an aid in working out the acoustical design of an auditorium, scaling techniques have recently been developed.[28] A model of the proposed auditorium is built to the scale of, say, one-tenth the size of the original prototype. Sounds of short wavelengths produced inside this model will behave in the same way as sounds of ten times the wavelength in the prototype, provided that corresponding reflecting surfaces have the same absorption coefficients. By investigating the behavior of the model, deficiencies may be uncovered and corrective measures tried out at considerably less expense than would be required in the auditorium itself.

Practice and performance rooms for music schools also require considerable care in planning to insure proper reverberation times. It is also very important that they be provided with sufficient sound insulation from one another, and from adjoining areas. Criteria for such rooms have been well established.[29] In fact, any room in which sound is to be produced will benefit by some acoustical considerations; this applies to conference rooms, schoolrooms, and even rooms for cocktail parties.[30]

Sound Amplification

For large auditoriums, as well as for outdoor performances, electronic amplification (to be discussed further in Ch. 15) is usually necessary, and it is obvious that the design of such an amplifying system must be worked out in conjunction with the acoustical design of the auditorium.[31] Any amplification used must be unobtrusive; the listener should not be aware that the sound from the stage is being amplified. This requires high-quality equipment to reproduce the sound with fidelity.

In sound-amplifying systems, a common arrangement is to mount the loudspeaker producing the amplified sound directly over the stage, so that sound reaches the listener from the stage and from the loudspeaker at about the same time. However, sometimes it is necessary to mount the loudspeakers out over the audience. If this is done, an interesting *precedence effect* occurs which must be taken into account.[32] If a listener hears a source of sound on a stage some distance off, and at the same time hears the same sound amplified and coming from a loudspeaker close by, he will judge that all of the sound comes from the loudspeaker. This is because electrical signals travel along wires much faster than sound travels through air, so that the listener hears the sound from the loudspeaker before he hears that from the stage, and his ear decides that all the sound comes from the source it hears first. However, if the electrical signal is delayed in some fashion so that the sound from the loudspeaker reaches the listener later than some 50 milliseconds after the direct sound from the stage, the listener will judge that all of the sound comes from the stage, even though that from the loudspeaker may be considerably stronger. Proper recognition of the precedence effect will obviously be necessary if realistic and unobtrusive sound amplification is to be obtained. Commercial devices are now available which will produce delays of the proper amounts by means of electrical circuits.

The use of sound amplification introduces other problems. One of the commonest of these is the so-called *ringing* that can occur when the microphone attached to the amplifying system picks up sound from its own loudspeakers. The result is usually a loud and unpleasant sound of steady pitch. An amplifying system that is on the verge of ringing will produce considerable distortion of the reproduced sound, even though no actual ringing sound occurs. Recent work has resulted in the development of systems for suppressing the tendency to ring; these can produce a great improvement in the performance of amplifying systems.[33]

There remains one aspect of auditorium acoustics about which very little has been done: this is to determine how the acoustical environment affects the performer on the stage. For example, it is important for a musician in an ensemble to be able to hear the others in the ensemble; if the musicians cannot hear each other, the conductor will find it more difficult to keep them together. This *ensemble* factor is frequently overlooked. Also, too much absorption in the auditorium makes it difficult for the musician to produce a satisfactory sound level, as we have already noted. In this situation the discreet use of electronic amplification can be of assistance to the performer.[34] Unfortunately, such amplification is very often overdone. Much more research is needed to determine the relationship between the musician and his acoustic environment.

PART IV

THE PRODUCTION OF MUSICAL SOUNDS: MUSICAL INSTRUMENTS

CHAPTER TEN

THE STRING
INSTRUMENTS

Now that we have discussed various vibrating systems, the sound waves they produce, and the reception of these waves by the ear, we can turn our attention to musical instruments as the practical sources of musical sounds. Their acoustical behavior is quite complicated and by no means thoroughly understood. As a consequence much of what we will say about musical instruments in the following pages is true now to the best of our knowledge, but is subject to revision as we learn more about them.

The vibrating string discussed in Ch. 4 is one of the simpler systems having more than one resonance frequency and serves as the acoustical foundation for the family of string instruments; this family is one of the most important in music. Reduced to its fundamentals, a string instrument is a hollow wooden box that serves to support a number of strings and maintain them under tension. The strings are plucked or bowed, producing the complex vibrations described in Ch. 4. These vibrations are transmitted through the string supports to the wooden box, causing its various surfaces to oscillate; these then produce vibrations inside the box and in the surrounding air that are audible as sound. However, as is frequently the case in music, the proper musical accomplishment of this objective is very difficult. Years of cut-and-try experimentation by violin makers have developed methods of building string instruments that are reasonably satisfactory, although unpredictably so, and have produced splendid models; but a good acoustical basis for the rational construction of string instruments has remained largely unknown until recent years. As a result of acoustical investigations extending over a considerable number of years and still going on, the principles underlying the design of string instruments are now fairly well known; however, the proper application of these principles to the actual construction of instruments is still full of unsolved problems.

The family of strings has a number of members of varying importance. Some, like the guitar, are plucked. In some the strings are struck by hammers. The piano is a member of this group, but it is so important that we will give it separate consideration. The instruments of the orchestral string family—violin, viola, violoncello, and double bass—are sounded by plucking or bowing, with the latter method predominating. The members of this family have much the same acoustical problems in common, so we may take the violin as representative.

Construction

The *violin* is sketched briefly in Fig. 1. The strings whose vibrations are the original source of the sound are fastened to the *tail piece* at one end, pass over the *bridge* mounted on the body of the instrument, and are wound around *pegs* situated in the *peg box* at the other end. Turning a peg adjusts the tension of the string and tunes it to the proper fundamental frequency. For most of their length the strings pass over

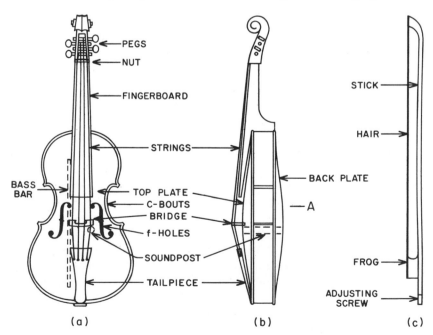

(a)   (b)   (c)

Fig. 1. (a) Front view of a violin. (b) Side view. (c) Bow.

the *fingerboard.* The vibrating length of the string for the lowest fundamental (open string) is defined by the bridge at one end and the *nut*
at the other; pressing the string against the fingerboard with the finger
shortens the string and raises its frequency. Four strings are provided,
one for each finger in the left hand; they are tuned to G_3, D_4, A_4, and
E_5. Those for the lower frequencies are generally wound with wire to
increase their mass. This provides for vibrations of lower frequency
without having to make the strings too thick and hence too stiff and
without having to reduce their tension too much; both these extremes
would adversely affect the tone. The strings are set into vibration by
drawing the bow across them; the narrow center portion of the violin
produced by the graceful in-curves of the so-called C-bouts allows the
outside strings to be bowed without scraping the bow against the sides
of the violin.

Action of the Bow

The string instruments are members of the large group of musical
instruments which can produce a steady tone. This tone was defined
in Ch. 6 as periodic, having a vibration whose waveform is complex
but does not change appreciably from one cycle to the next. The violin
string is set into such steady vibration by drawing across it a properly
constructed *bow;* this is a simple operation, but one whose smooth
accomplishment is a result of a great deal of practice on the part of the
string player. The violin bow is made by fastening one end of a bundle
of strands of horsehair to one end of a thin shaft of wood. The other
end of the bundle is attached to a piece called the *frog,* mounted on the
other end of the stick through a screw arrangement that allows the
position of the frog to be varied. Doing this adjusts the tension of the
horsehair to the proper value. The bowing is done on the string usually
about halfway between the bridge and the end of the fingerboard.

The bow is placed on the string, at right angles to it, and is moved
back and forth across it. The bow then pulls the string with it because
of the friction between the horsehair and the string; this friction is
generally made as large as possible by rubbing rosin on the horsehair.
As the string is displaced to one side, the tension in it results in a restoring force tending to pull the string back to its original position, and
this force increases as the string is pulled farther away. The forces acting under these conditions are shown in Fig. 2. The frictional force of
the bow on the string has a largest possible value, as explained in
Ch. 1. As the displacement of the string increases, the restoring force
gets large enough to overcome the frictional force, and the string starts
to slide under the bow and move back toward its normal position.
When the string is moving across the bow, the frictional force becomes

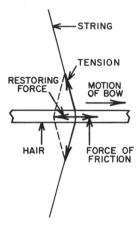

FIG. 2. Forces acting during bowing of a string.

smaller; this is a fairly general property of frictional forces between surfaces that are not lubricated. As a result, once the string starts moving, it can go some distance before the restoring force becomes small enough to allow the bow to take hold of the string again. Furthermore, the inertia of the string helps to keep it moving and carries it to some distance past the starting point. The bow finally catches the string again, pulls it again to one side, and the cycle is repeated. The motion of the bow will be very nearly uniform during one cycle of this vibration, so the resultant motions of the string and bow will be as plotted in the graph of Fig. 3. That portion of the string in contact with the bow moves with it during most of the cycle, and then suddenly snaps back across the bow to a new position on it, and this process is repeated periodically.[1]

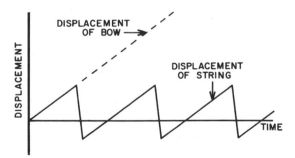

FIG. 3. Graph of the displacement of the violin bow and the displacement of the string, plotted against time.

The motion of the string as produced by the bow is thus far from being sinusoidal; rather, its motion at the point of bowing is almost the sawtooth wave described in Ch. 6. It is not quite the same because the string does not snap back instantly to its original position, but instead requires a certain fraction of the cycle. The motion of the string is thus a complex vibration containing all the harmonics of the fundamental. The relative amplitudes of these harmonics will depend in a complicated way on the spot on the string where it is bowed, on the speed of the bow across the string, and on the force the bow exerts on the string. (This last is sometimes called the *bowing pressure*, although actually it is a force, not a pressure.) These harmonics of the string vibration will appear in the radiated sound, giving it a specific quality. If the string is bowed closer to the end of the string at the bridge (*sul ponticello*), the proportion of high harmonics is generally increased and the tone is "brighter." Conversely, if the string is bowed further down toward the fingerboard (*sul tasto*), the proportion of higher harmonics is reduced and the sound is of a softer quality. A player thus has some measure of control over the quality of the sound his instrument produces.

The amplitude of the string vibration at a given bowing point will depend on the speed with which the bow is drawn across the string, and on the distance of the bow from the bridge. These will partially determine the loudness of the tone. The vibration amplitude and hence the loudness is independent of the bowing pressure.[2] The frequency of the tone is determined by the mass of the string, its tension as adjusted by the peg, and the length of the string between the bridge and the point where the player's finger presses it to the fingerboard. The frequency does not depend on the bowing pressure as long as there is enough to maintain the contact between the bow and the string.

It is possible for a string to vibrate longitudinally in the same fashion as the bar discussed in Ch. 4; this vibration can be produced by drawing the bow along the string rather than across it. The frequency of this vibration is independent of the tension of the string, and it is generally badly out of tune with the usual transverse vibrations. When it occurs excessively (as a result of unskilled playing) it is a very unpleasant sound. The small amount of longitudinal vibration produced in normal playing is not transmitted to the violin body by the bridge.[3]

The violin *mute* is a small device that can be affixed to the top of the violin bridge. One model clips onto the bridge by means of prongs fitting between the strings. Another model fits onto the strings between the bridge and the tailpiece. In normal playing it is away from the bridge, and it is put into use by sliding it along the strings until it rests on top of the bridge. With either kind, placing the mute in position increases the mass of the bridge; the sidewise forces of the vibrating strings will then produce smaller accelerations and hence smaller amplitude vibrations of the bridge. As a consequence, there will be a reduction in amplitude of the harmonics in the string tone, and the amount of reduction is greater the higher the frequencies of the harmonics. Hence the tone of the instrument with the mute in place is softer and of a less bright quality than without the mute.

The Violin Body

The hollow box that forms the body of the violin serves two functions: it supports the strings so they can vibrate properly, and it makes the sound from the strings audible. The vibrating strings themselves disturb very little air and hence radiate practically no sound; the important job of the violin body is to transmit these string vibrations to the air in such a way as to give them the proper loudness and tone. The quality of the sound the violin radiates will depend not only on the harmonic structure of the vibration of the string, but also on the way in which these harmonics are transmitted through the bridge to the body of the violin and thence to the air.

The essential parts of the body of the violin are shown in Figs. 1 and 4. Fig. 4(a) is a lateral section taken at the point marked A in Fig. 1, and Fig. 4(b) is a longitudinal section through the center.

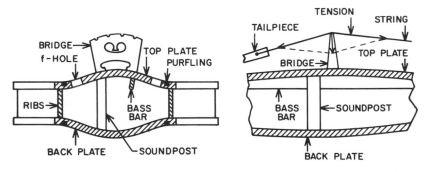

FIG. 4. (a) Lateral section of a violin, taken at the point marked A in Fig. 1. The plate thicknesses are much exaggerated. (b) Longitudinal section, showing forces exerted by strings on the bridge.

The violin box, of elaborate shape, consists of two thin plates of wood: the *back plate,* the side away from the strings; and the *top plate* or *belly,* the side adjacent to the strings. These plates are glued to strips of wood called *ribs,* which form the sides of the box and give the instrument its characteristic shape. The air space inside the box communicates to the outside through the *f-holes* cut in the top plate. The bridge carrying the four strings is mounted on the top plate midway between the f-holes. The (unbowed) strings passing over the bridge exert a downward force on it that is transmitted to the top plate as shown in Fig. 4. The combined tension of the strings is approximately 50 lbs. force, and the downward force is some 20 lbs.[4] A strip of wood called the *bass-bar* is glued longitudinally on the inside of the top plate under one foot of the bridge; it helps to stiffen the plate against this downward force. Inside the box, nearly under the other foot of the bridge, is placed a short wooden stick called the *sound post,* which extends from the top plate to the back plate and is held in place by friction; it serves to make a rigid connection between the two plates and its position is quite critical. The sound post is a very important coupling point between the top and back plates, transferring vibrations from one to the other.

When a string is bowed, its vibrations produce sidewise forces on the bridge. These forces are not shown in Fig. 4; they are directed perpendicular to the plane of the paper and cause the bridge to vibrate in its own plane. These vibrations are then transmitted to the violin box.

It is the vibrations of the box of the violin as produced by the strings acting through the bridge that determine how the instrument will sound. The wooden plates of which it is made have a complicated vibration spectrum. Fig. 5 shows resonance curves for the top and back

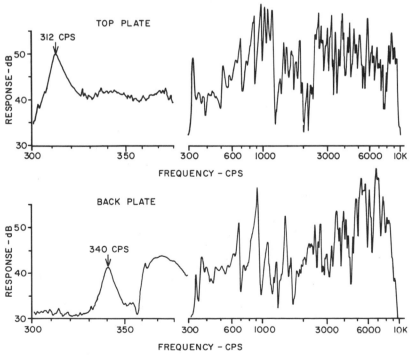

FIG. 5. Resonance curves of the top plate and the back plate of a violin. The lowest resonance of each plate is shown on an expanded frequency scale.

plates of a violin; these curves were made as described below. Each plate has very many resonance frequencies. The box, being a combination of these two plates, will then have as complicated a spectrum of resonances; this may be seen in Fig. 8 below. In addition, the hollow interior communicating to the outside through the f-holes forms a Helmholtz resonator (as described in Ch. 4) whose resonance frequency depends on the volume of the box, the size of the f-holes, and the thickness of the edges of the f-holes. This resonance is called the *air resonance*; in contrast to the box, there is only one measurable air resonance.[5]

As might be expected, the positions and spacings of these resonances are crucial in determining how the violin will sound. The rule-of-thumb methods employed by the violin makers were such as to put these resonances in approximately the right places, although the artisans were unaware that this was what was being accomplished. Recent investigations involving the measurements of resonance frequencies on a great many violins have developed the acoustical basis for the best placing of these resonances.[6]

To begin with, it is found that the air resonance is the lowest of all. Furthermore, the best results are obtained when the dimensions of the violin body and the size of the f-holes are such that the frequency of the air resonance coincides approximately with that of the D string, seven semitones above the frequency of the lowest string. If this is done, the air resonance reinforces all vibration frequencies in the neighborhood of D_4. When this resonance is excited, the air in the box is vibrating through the f-holes; however, measurements show that most of the sound produced is not radiated from the f-holes, but is instead radiated from the top and back plates of the wooden box, which are set in motion by the varying pressure of the air inside the box.[7]

Of the many resonance frequencies possessed by the box, there will be one lowest one, called the *main wood resonance*. In a good instrument, this resonance should be approximately a fifth higher than the air resonance, to reinforce tones in the vicinity of A_4. This includes not only the lowest tones on the A string; just as important, it will reinforce second harmonics of tones in the neighborhood of A_3, an octave lower, and hence will emphasize the low notes on the G string. This is because increasing the amplitudes of the harmonics of a complex tone gives the auditory illusion of an increase in loudness of the fundamental, even though the fundamental itself has not been changed.

For the same reason, the higher vibration modes of the box, of which there are a great many, will reinforce not only the notes whose fundamentals are near these resonance frequencies, but also harmonics in this range. Thus a resonance frequency near D_6 will reinforce a fundamental of this frequency; it will also reinforce the second harmonic of D_5, whose fundamental is half the frequency; likewise the third harmonic of G_4, of fundamental one-third the frequency; the sixth harmonic of G_3, of fundamental one-sixth the frequency, and so on. The presence of these higher resonance frequencies in the violin box is thus quite essential to the reinforcement of lower frequencies.

Tap Tones and Plate Resonances

To construct a box of this sort, one that will reinforce all the harmonics of the string vibration in the most musically desirable way, is not an easy task. The present size, shape, and constructional details of the violin body have been arrived at through many years of experimentation, most of it done about three hundred years ago. The frequency of the air resonance depends on the dimensions of the box and the size of the f-holes, and can be set to a given value without too much trouble. The same is not true of the wood resonances; their frequencies depend not only on the dimensions of the violin box, but also on the elastic properties of the wood from which the box is made. The tone of a violin hence depends to a great extent on the wood from which it is constructed. This is in contrast to the woodwinds, whose tone does not depend on the material; we will discuss this in Ch. 11.

Wood is a very unpredictable material: every piece is different from every other piece. For this reason, it is impossible to prescribe in advance the thickness of the top and back plates that will give the desired resonance frequencies to the violin body. Each piece of wood being made into a plate must be individually worked down to the right thickness. To determine how far to go in the thinning-down process, the violin maker has customarily employed *tap tones*; he holds the plate near one end between thumb and forefinger, taps it at various points, and listens to the resulting sounds. He works to achieve a clear bell-like tone in each plate as well as to obtain the desired pitch of each, and from these sounds he determines whether or not the plates are right. The ability to judge the proper relation of the tap tones of the top and back plates is an important part of the art of violin making.

Because of the complicated form of the plates, there will usually be several resonances below 600 cycles per second, the frequency region in which the tap tones lie. This makes it quite difficult to judge by ear the frequencies of these tones. However, recent investigations have considerably refined the analysis of tap tones by providing an accurate means of measuring the resonance frequencies of the plates.[8] The method is similar to the external excitation method of measuring resonances that we described in Ch. 4. The plate is suspended by means of rubber bands attached to its corners, so that it is free to vibrate. A very light transducer (these will be discussed in Ch. 15) is attached to the center of the plate; this device will produce vibrations in the plate when an electrical driving signal is supplied to it, the vibration frequency being determined by the frequency of the signal. The amplitude of the plate vibration is then observed as the driving signal frequency is changed. This may be done by measuring the sound produced by the plate at some chosen distance as indicated by a sound level meter such as was described in Ch. 5. With a constant amplitude driving signal, the plate will vibrate strongly at its resonance frequencies, and these will appear as higher intensities (peaks) in the measured sound.

In this way we may determine the resonance frequencies of the top and back plates; these may later be correlated with the sound of the instrument when the plates are assembled. This has been done for many instruments, and the overall result for violins is that the average of the top plate and back plate lowest resonance frequencies is about seven semitones below the main wood resonance frequency of the finished instrument. (However, we cannot say that the plate tap tones have moved up seven semitones when the instrument is assembled; gluing the plates to the ribs produces a quite different vibrating system.) This fact enables the violin maker to adjust the plates so as to put the main wood resonance in its proper place, and it eliminates a great deal of guesswork from the art of violinmaking. It has been found further that the violin sounds best when the lowest resonance frequencies of each plate differ by between a tone and a semitone, with the back plate at a higher frequency than the top. The frequencies must not coincide.

Fig. 5 shows resonance curves for the top and back plates of a violin, made by the excitation method described above. At the left side of Fig. 5, the first resonances for each plate are shown on an expanded frequency scale so as to exhibit them more clearly. The resonance frequencies differ by 1.5 semitones, with the back higher than the top, so the two plates fit the criterion outlined above.

Loudness Curves and Response Curves

If a violin is bowed as strongly as possible (without vibrato) and its sound output measured on a sound level meter for each of the notes of the chromatic scale, a curve of sound output vs. frequency may be plotted. Such a curve is called a *loudness curve*.

Loudness curves are useful in giving an overall picture of the behavior of the instrument. Fig. 6 shows loudness curves for two violins. The top curve (a) is for a good 1713 Stradivarius violin and shows the

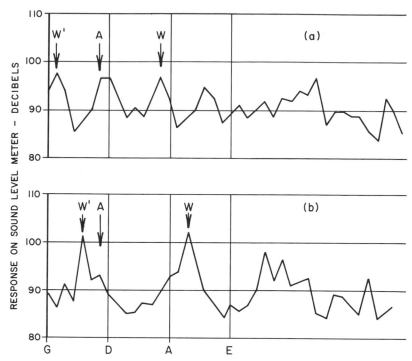

Fig. 6. Loudness curves for two violins. (a) A good 1713 Stradivarius violin. (b) A 250-year-old violin of unknown origin. (Reproduced from ref. 4. Copyright © 1962 by Scientific American, Inc. All rights reserved.)

positions of the two main resonances relative to the frequencies of the strings, which are shown on the bottom of the graph. The resonance marked W is the main wood resonance, and A is the air resonance; they coincide very nearly with the frequencies of the A and D strings. The resonance marked W' (dubbed "wood-prime" by the people doing this work) is an octave below W and is produced by the resonance of W with the second harmonic of frequencies near the G string, as explained above. Curve (b) of Fig. 6 is for a 250-year-old violin of unknown origin. In this instrument the wood resonance is a little too high in frequency and the air resonance not only too low, but weak as well. Hence there is a wide region in which the response is poor, and the tone of the instrument suffers.

Response curves are made by an electrical method similar to that described above for the measurement of plate resonances.[9] A transducer is mounted on the bridge of the violin, excited by a variable frequency electrical signal, and the sound output of the instrument measured with a sound level meter. This method requires more equipment than the method of simply bowing the instrument, but it has the advantage of showing the resonances in considerable detail. Fig. 7 shows curves made by both methods for a viola, the top curve (a)

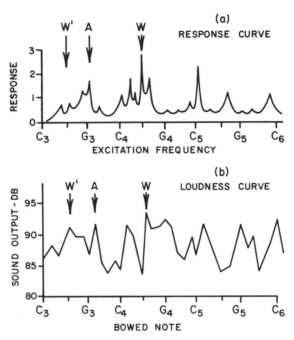

FIG. 7. Response curve and loudness curve for a viola.

being made by the electrical method.[10] The main wood resonance and air resonance are evident; however, the W' peak which appears in the bowing curve (b) is missing in the top curve (a). This is because the electrical signal has no harmonics above the fundamental, so that frequency W' will produce no response in the resonance W an octave higher.

Another example of a response curve is shown in Fig. 8, the curve for a violin assembled from the two plates whose resonance curves are given in Fig. 5. This curve was made with more sensitive equipment than was used to obtain the response curve of Fig. 7(a), and so shows more detail. It was made with the strings on the instrument, so their resonances are added in with those of the violin body. The overall spectrum is obviously quite complex. The air resonance and main wood resonance are not specifically identified, but are among the first half-dozen peaks on the left side of the curve. Note that the resonances

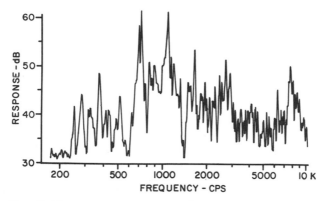

FIG. 8. Response curve for a violin assembled from the two plates whose resonance curves are given in Fig. 5.

above about 2000 cycles per second are much less pronounced in the assembled instrument than they are in the individual plates; this is apparently a desirable condition.

Response curves and loudness curves have been made on many violins, ranging in quality from excellent Stradivarius fiddles to a $5.00 instrument used as a "standard of badness," to see to what further extent response curves can be correlated with the quality of the instrument. It appears that a good violin is one whose response curve is relatively uniform throughout its useable frequency range whereas a poor one will show frequency regions where the response is too large or too small. Further work in this direction is continuing and eventually it should be possible to determine from the response curves of the individual plates how the assembled instrument will behave, and to judge accurately from the response curve of the instrument how it will sound when played. In time, it may be possible to learn how to shape the plates to get the desired response curves. However, a great deal of further research is necessary to correlate response curves with violin quality.

Miscellaneous Factors in Violin Tone

A number of other matters have been investigated to see what bearing they may have on the tone quality of the violin. For example, the excellent quality of Stradivarius violins has frequently been attributed to a varnish of almost magical properties used by the old instrument makers, whose secret has been lost. However, resonance curves made on instruments in the varnished and unvarnished condition show that, as normally used, the varnish itself has rather little effect on the tone.[11] Too much or too hard a varnish can have adverse effects. However, the main function of the varnish appears to be to seal the pores of the wood to keep out dust, which otherwise would adversely affect the properties of the wood. The effect of varnish is still being studied.[12]

The question of aging is being looked into; it does not appear that an old violin is better than a new one simply because it is older. Some changes of the properties of wood with age appear to improve violin tone; these are being investigated. However, age is not the only factor.

The various properties of woods used in violin making have been measured.[13] The values of density, elasticity, and damping that appear to be best have been worked out, but other properties still remain to be evaluated. If the various properties of woods that are important to violin tone can be isolated and measured, it should be possible to develop synthetic materials with these properties from which to manufacture violins. (Whether string players would accept instruments made of synthetic materials is another question.)

It does appear that playing a new violin can improve its tone. One reason is as follows: the vibrations of the top plate of the violin are affected by the fact that it is glued firmly to the ribs, and it has been found that the tone of the violin is better if the joint between top plate and ribs is relatively flexible. As an example, it was found that a cheap violin could have its tone improved and power increased considerably by cutting a narrow groove around the outside edge of the top plate; this would allow for more bending at this joint.[14] The so-called *purfling* of the violin is just such a narrow groove cut around the outside edge of the top plate and filled with thin strips of wood that are glued in; it is shown in Fig. 4(a). When new, the joint between the top and ribs is relatively stiff, but as the violin is played the glue holding the strips of wood in the groove apparently cracks loose to some extent and allows the joint more flexibility. After a period of playing, the purfling becomes sufficiently loosened (but not so loose as to fall out) and the tone of the violin is better.

There are other reasons why the tone of a violin is improved by playing it. These have to do with such things as changes in the wood structure that occur when the wood is stretched and compressed by continual vibrations. These matters still need much investigation.

The vibrato, such a characteristic part of violin playing, has been investigated and found to be a fluctuation in both amplitude and frequency. The amplitude variation is not as important as the frequency variation, which is what gives the tone its characteristic quality. This frequency variation is to be expected since it is produced by moving the finger back and forth on the string and so periodically varying its length. The frequency of the vibrato (as distinguished from the frequency of the string) is some five to six vibrations per second; the frequency of the string varies over a range of some 30 to 45 cents.[15]

The sound output of the violin can be varied by the player over a range of about 30 decibels.[16] The total output is not large; a number of violins are required to give a sound output equivalent to that of a single brass instrument. In addition to the louder tone, a group of violins playing together provides a richer tone through the chorus effect discussed in Ch. 6; this tone is the mainstay of the symphony orchestra.

Listening tests have been made of the playing of old Italian violins as compared to good modern violins under "blindfold" conditions. These are conditions under which the listeners do not know, in any particular instance, which of the group of instruments under comparison is being played. Such test conditions are necessary for any valid judgment to be made, since musicians, like other people, are prone to hear what they want to hear. Listening tests made under such conditions show that most string players cannot distinguish between the sound of an old Italian violin and that of a good modern instrument.[17] A very few individuals are able to make this distinction under these conditions, so that one cannot unequivocally say there is no difference between old and new violins. However, the string player with sufficient strength of character to withstand the temptation of the status implied in owning a Stradivarius or Guarnerius, or with insufficient funds to yield to such temptation, can find a modern instrument that will be perfectly satisfactory in its tone and power.

The Larger String Instruments: Scaling

The larger string instruments are the *viola, violoncello,* and *double bass.* The general shape and construction of the first two is similar to that of the violin. The strings are tuned a fifth apart in the viola and cello— C_3, G_3, D_4, A_4 in the viola; C_2, G_2, D_3, A_3 in the cello. The double bass is a member of the viol family; it is shaped and constructed somewhat differently than the violin. In the double bass the fingerboard becomes quite long, so the strings are tuned a fourth apart— E_1, A_1, D_2, G_2—to keep down the finger spacings and the distances the hand must travel.

The larger size of these instruments introduces more problems. It can be demonstrated mathematically that if we want to lower the frequency of an instrument by a given factor, we might do so by increasing all the dimensions of the instrument by that factor; this process will be called *linear enlargement.* For example, suppose we want a viola to play a fifth lower than the violin for corresponding finger positions; all frequencies will then be $\frac{2}{3}$ the corresponding one for the violin, as we saw in Ch. 8. We could do this by making all the dimensions of the viola $\frac{3}{2}$ the corresponding ones in the violin; the viola would thus be half again as long as the violin. Similarly, a cello, to play an octave and a fifth lower than the violin (one-third the frequency), would have to be three times as long. All other dimensions would have to be increased similarly, so that the thickness of wood in the cello would be three times as great as for the violin, the width of the cello would be three times as great, and so on. The materials should also be identical. If all these conditions are met, the larger instrument would theoretically have all its resonances in the correct places and its behavior would be the same as the smaller instrument at corresponding tone ranges.

Such simple linear enlargement to produce the larger instruments is not possible; the dimensions do not turn out to be practical. A viola fifty percent longer than the violin cannot be played in the usual manner, and a cello three times as large as a violin is much too awkward. Furthermore, materials will never be identical, since wood as a body material is such a variable substance; for this reason it is even impossible to make two violins of identical dimensions that are identical in quality.

In practice, therefore, considerable compromise must be made in the sizes of the larger instruments. The viola is actually built only some 15 to 20 percent longer than the violin and a cello is a little more than twice as large, so that the instruments are of playable size. Fig. 9 shows a comparison of the dimensions actually used with those that would be given by linear enlargement.[18] The violin is taken as standard, and is shown by a solid black circle. The frequency relative to the violin is plotted horizontally. The dotted diagonal line gives the size of the instrument relative to the violin on the basis of simple linear enlargement; it shows, for example, that an instrument of one third the frequency should have three times the size of the violin. The actual sizes of the viola and cello are shown by the open circles; they depart considerably from the diagonal line representing linear enlargement.

This departure has an unfortunate consequence; it means that the air resonance and the main wood resonance in the viola and cello will be at the wrong frequencies to coincide with those of the open second and third strings, as they do in the violin. If nothing were done to compensate, they would come out too high. The air resonance in a cello of just twice the dimensions of a violin would come out to be an octave lower than the violin, or near D_3 instead of near G_2 where it belongs. As a result, the lower tones of the viola and cello would suffer through not being sufficiently reinforced by the body.

Changing the positions of the resonances to put them in the right places calls for a rather complete remodeling of the instrument. The old Italian masters who perfected the violin apparently did not work enough on the larger instruments to develop the designs that would put the resonances in the right places, so the traditional viola, cello, and double bass are still somewhat deficient in this respect. However,

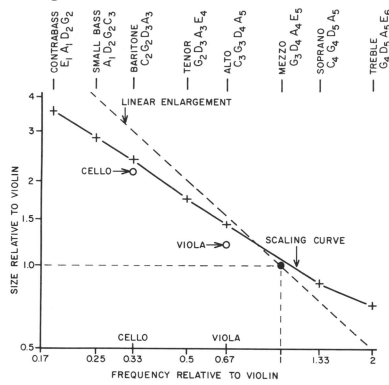

Fig. 9. Scaling curve for string instruments as compared to simple linear enlargement, together with names and tunings of new instruments built according to scaling principles.

as a result of recent research, the principles of design for the larger instruments which will place the two main resonances at the two middle open strings have now been worked out reasonably well. The procedure for determining overall dimensions, plate thicknesses, elastic constants, and so on to put the resonances in the proper positions is called *scaling*.[19] In Fig. 9 the crosses connected by solid lines give the *scaling curve*, used at present to determine the size, relative to the violin, of an instrument designed to play at a given frequency relative to the violin. According to this curve, for example, an instrument designed to play an octave lower than the violin would have 1.7 times its length. Other quantities such as plate thicknesses, rib heights, and so forth must then be worked out from the complete scaling theory.

The practical results of the application of the principles of scaling to the construction of string instruments are embodied in the design of a viola and cello with properly placed resonances. In addition to these, a number of instruments of other sizes have been designed and built along the same principles, and the result is a whole family of new string instruments.[20] These range from a small treble violin, tuned an octave above the conventional instrument, down to a seven-foot double bass. The names and tunings of these instruments are shown in Fig. 9; their scalings are shown by the crosses directly under the names. These instruments have been demonstrated in several public concerts and have aroused considerable enthusiasm among the string players who have heard and played them. The viola in particular has an excellent tone, and the large bass is exceptionally powerful. (This is the opinion of the author, who has heard them played.) This group of instruments is a good demonstration of the practical benefits that can be realized from painstaking investigation of the acoustical behavior of musical instruments.

The "Wolf" Tone

The main wood resonance in a string instrument can be set into vigorous vibration by bowing a string at the frequency of this resonance. The amplitude of this vibration can become so large that the body of the instrument no longer serves as an adequate support for the string. When this happens, the body resonance tries to determine the overall vibration frequency, which normally is determined only by the string. The result is a conflict between the two, which appears as a beat between the two frequencies; this produces a poor and unsteady tone called the "wolf" tone. It is of frequent occurrence in the cello. The main body resonance is usually quite narrow, so that its response to a string vibration diminishes rapidly as the string frequency gets further away from the resonance frequency. As a result, if the wolf note lies midway between two semitones on the instrument, it will generally not cause too much trouble. If it lies too close to one of the scale frequencies, then various means must be used to "tame" it; these will involve changing its frequency by cut-and-try alterations such as devices fastened to the tailpiece and applying adjustable forces to the belly through pads, etc.[21] The wolf tone cannot be completely killed since this would reduce the ability of the instrument to radiate sound in a very important part of its range.

It should be obvious from the brief discussion of this chapter that a great deal still needs to be learned about the acoustical behavior of the string instruments. The acoustical knowledge outlined above has been the result of the work of a relatively few people; it is hoped that more individuals with an interest in acoustics and music can participate. An organization incorporated as the Catgut Acoustical Society has been formed to promote work in the strings; it should have the blessing of all.[22] Eventually, it should be possible to develop string instruments of fine tone at prices even students can afford.

CHAPTER ELEVEN

THE WOODWIND INSTRUMENTS, AND OTHERS

Just as the vibrating string is the acoustical basis for the family of string instruments considered in the previous chapter, so also are the vibrating air columns, discussed in Ch. 4, the acoustical foundations for the family of woodwind instruments. The open cylindrical tube (that is, open at both ends) becomes a flute. The cylindrical tube closed at one end becomes a clarinet. The conical tube is developed into the oboe and bassoon. As the instruments are actually constructed, the shapes of their air columns deviate somewhat from simple cylinders and cones, and these deviations turn out to be of considerable musical importance.

In all the wind instruments a longitudinal standing wave is generated and maintained in the air column inside the instrument. Openings in the instrument wall define the length of the standing wave and allow it to radiate the sound that is heard. To generate this standing wave requires power, and there are two practical ways in which acoustical power is supplied to the wind instruments to maintain the vibrations of the air column. The flute uses an oscillating air stream; the reed instruments, as the name implies, use the mechanical vibrations of a thin piece of elastic material called a *reed*. The two systems differ considerably in the way they function. That for the flute is simple in appearance, but this simplicity is deceptive and conceals a number of acoustical problems; however, let us consider it first.

Edge Tones and Air Column Vibrations

Nature sometimes seems to enjoy adding complexities to systems that one would think should be basically simple. The motions of fluids are a good illustration of this annoying tendency. Suppose we have a tube that terminates in a relatively narrow slit and which is connected to a supply of air under pressure, so that a stream of air emerges from the slit. Opposite the slit and a short distance away is a sharp edge so placed that the air stream will strike it; the arrangement of slit and edge is as illustrated in Fig. 1. With any reasonable pressure inside

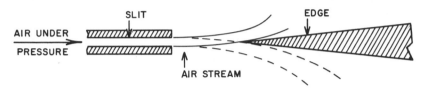

Fig. 1. An arrangement for producing edge tones.

the box, the air stream does not simply split in two when it strikes the edge, as one would expect a simple and well-behaved system to do; instead, the stream vibrates back and forth across the edge, as indicated by the solid and dotted outlines in Fig. 1. This vibration is usually in the audible frequency range, and the system then becomes a source of sound waves. Sounds produced by this mechanism are called *edge tones.*

The detailed behavior of edge tone systems is quite complex, but a considerable amount of work has been done on them and they are now beginning to be understood.[1] Their frequency depends in a complicated manner on the distance from the slit to the sharp edge and on the speed of the air emerging from the slit. In particular, for a fixed slit-to-edge distance, the vibration frequency will increase as the speed of the air stream is increased.

For musical purposes, the edge tone system is always attached to a tube or pipe having an air column, as shown in the cross-sectional sketch in Fig. 2. With this arrangement, blowing air through the slit

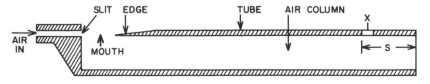

Fig. 2. Edge-tone system connected to an air column.

will produce steady oscillations in the air column at one of its resonance frequencies rather than at one of the edge tone frequencies. This happens in the following way: When the air is first blown through the slit, edge tone vibrations begin to occur. Puffs of air made by the air stream vibrating across the edge enter the air column and produce vibrations of some of its resonance modes, just as plucking a string excites its resonances. One of these, usually the fundamental, will build up most rapidly, and air will begin flowing in and out of the gap between the slit and the edge at the frequency of this mode. The gap between the slit and the edge is called the *mouth* of the pipe. When air flows into the pipe through the mouth, it pulls the stream from the slit with it. The air stream then blows into the pipe at the same time that the air in the column at the mouth is moving down the pipe. (The term *down* as used in connection with the woodwinds means away from the end in which air is blown.) Conversely, when the air in the vibrating air column flows out of the mouth, it pushes the air stream from the slit out away from the edge, so that the air stream is also directed out of the mouth. The air stream thus aids the air in the pipe in its motion, and the amplitude of the vibration increases. A standing wave then builds up in the pipe; this wave has a displacement antinode at the mouth since the mouth behaves like an open end.

The behavior of this acoustical system resembles that of the mechanical mass and spring system of Ch. 2 which was given a small push in the direction of motion each time it passed through the equilibrium position. Energy was then being supplied to the system, and the amplitude of the vibration built up until the energy lost during most of a cycle was made up by energy gained during the short part of the cycle that the push was applied. In the same way, the oscillations in the pipe produced by the vibrating air stream build up until the energy lost by friction of the moving air column against the walls of the pipe and by sound radiation during a cycle equals the energy supplied by the air stream during a cycle. The power supplied by the air stream is then equal to the power dissipated by the standing wave, and the amplitude of the vibration remains constant.

A simple demonstration of a vibration maintained by the edge tone mechanism is afforded by blowing across an empty or partially empty bottle. The stream of air from the lips striking the edge of the bottle opening supplies the edge tone system, and the bottle acts as a Helmholtz resonator whose frequency is determined by its volume and by the size of the opening. The *whistle* as used by referees at sporting events is an adaptation of the same idea.

When a pipe is made to vibrate in this way, the frequency of the system is very nearly that of the fundamental frequency of the pipe.

If the speed of the air stream is increased by somewhat harder blowing, the frequency will rise, but rather little. The edge tone, which normally would rise in frequency with harder blowing, is not able to do so because the air column constrains it to vibrate at very nearly the column frequency. However, if the speed of the air stream is increased sufficiently, the vibration frequency jumps to that of the next higher resonance mode in the air column. For a tube open at the end away from the mouth, and so open at both ends, this will be approximately twice the fundamental frequency, so the tone produced will be an octave higher. The production of the higher resonance modes in the air column, as by blowing harder to increase the speed of the air stream, is called *overblowing*.

Tone Holes and the Basic Scale

The use of the edge tone system to generate musical sounds goes back to prehistoric times.[2] Many eons ago some early caveman found that by blowing in the right way across a hollow bone or stick he could produce a pleasing sound. The caveman also found that longer sticks and bones produced lower-pitched sounds. At some later time it was discovered that if holes were cut in the walls of the hollow stick and covered by the fingers, sounds of varying pitch could be produced by opening and closing the holes. Such holes are sometimes called *tone-holes* or *finger-holes*. Opening a hole in the side of the stick is equivalent to making its air column shorter, so the pitch would go up. If a hole is cut in the tube at the point X shown in Fig. 2, the tube behaves as if its length were shortened. The amount of shortening will depend on the size of the hole. It will vary from essentially nothing for a very small hole up to the distance S in Fig. 2 if the hole is large.

Over the centuries various adaptations of this system evolved. One is a simple primitive instrument made of bamboo which can be found in souvenir shops selling oriental goods. This instrument illustrates very well the fundamentals of woodwind behavior. A hollow section of bamboo is provided at one end with a plug, shaved off on one side to form a slit directed along the length of the stick; a notch cut in the side of the stick provides an edge upon which the air stream can impinge. The far end of the stick of bamboo is open. The result is just the system sketched in Fig. 2. A series of holes along one side of the bamboo tube, to be covered and uncovered with the fingers, allow various notes to be played. Six holes, three for each hand, are enough to provide the basic musical diatonic scale. With all the holes closed by the fingers, the fundamental of the tube produces the lowest note. By lifting the fingers in succession, starting with the one farthest from the blowing end, the diatonic scale may be played, one note at a time.

When all six holes are open, the seventh note of the scale will be reached. Now by closing all the holes again and overblowing, the octave of the original fundamental is produced. By again opening the holes in succession, the scale is traversed once more, but an octave higher. The instrument is now playing in a higher *register*; the point in the succession of fingerings where the register change occurs is called the *break*. This basic six-hole arrangement to provide the seven-note diatonic scale is found in all the woodwind instruments; we will call it the *basic scale*.

The Flute

The modern *flute* is a further development of the primitive instrument described above. A cross-sectional view of a portion of a flute (somewhat exaggerated in proportions, for clarity) is shown in Fig. 3. The air stream is blown across the tube instead of down its length. The air stream itself is produced by an orifice formed with the lips rather than one built into the instrument. This allows the player to alter the size and direction of flow of the air stream, and provides for more flexibility and control of the tone. The air stream is blown against the edge of a hole (the *mouth hole*) cut in a thickened area in the side of the tube, as shown in Fig. 3; the exact shape and contours of this hole are the result of much experimentation. Vibrations of the flute air column are then produced by the edge tone mechanism; the lips themselves do not vibrate, as flute players sometimes believe.

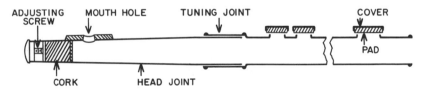

Fig. 3. Cross-sectional view of the essential parts of a flute.

Instead of using open holes to be covered by the fingers, all the holes of the flute are supplied with metal covers to be pressed down with the fingers either directly or through lever mechanisms. The covers have soft pads to provide a tight closure against the hole.

Flutes have been made of many materials, but wood and metal are the commonest; of this more later. (The term "woodwind" is not strictly accurate as applied to the metal flute, but is still traditionally used.) The body of the popular metal flute is made in three sections: the head joint, containing the mouth hole and a sliding cork for tuning purposes; the main body joint, with most of the key work; and a foot joint, with keys for the right-hand little finger. The joint between the head and the body is made moveable in order to tune the instrument over a small range. The air column of the instrument, called its *bore,* is cylindrical, about 1.9 centimeters (0.750 inches) in diameter, except in the head joint, where it is conical, narrowing to a diameter of about 1.7 centimeters (0.66 inches) at the end opposite the open end of the instrument. This taper, again the result of much experimentation, apparently compensates for some intonational problems in the higher register. The narrow end of the head joint is closed by a cork plug, shown in Fig. 3; the position of this plug can be changed slightly and adjusts the notes in the high register to be reasonably close to an octave above those in the low register. The overall length of the flute is about 67 centimeters (26.4 inches); the length from the mouth hole to the open end is about 60 centimeters (23.7 inches). An open tube of this length would have a lowest resonance corresponding to $C\sharp_4$ but, for reasons to be shown later, the flute sounds a semitone lower than this, blowing C_4 as its lowest note. The basic scale is the six-hole, seven-note scale described above, and it starts at D_4 with the six holes closed. Additional holes and keys are used to provide the sharps and flats and to extend the range down to C_4.

Smaller and larger models of the standard flute are built. The flute in G is about a third longer than the standard flute and sounds a fourth lower; the piccolo, which is about half as long, sounds an octave higher. Details of construction, keys, key mechanisms, and so on may be found in the literature.[3]

The Recorder

The *recorder* is an elaborated version of the primitive bamboo instrument described above, having the air stream produced by a slit built into the instrument and blowing along the tube instead of across it. Since the player cannot control the size and direction of the air stream, the instrument is less flexible than the flute. In some respects this makes it easier to play, and recorders have recently become quite popular. The present-day family consists of five sizes with fundamentals covering a two-octave range: bass recorder, F_3; tenor, C_4; treble, F_4; descant, C_5; and octave, F_5.

Reed Instruments: The Clarinet

Vibrating reeds have been used as sound sources since antiquity. Most all of us, in our younger days, have been shown how to produce a loud and satisfying blast of sound by blowing air past a blade of grass held in the proper manner between the hands. In this situation the blade of grass is acting as one simple form of vibrating reed. If such a reed is attached to an air column, it can maintain oscillations within the column at one of its resonance frequencies.

The *clarinet* is the simplest example of the production of vibrations in an air column by such a vibrating reed. A cylindrical tube, open at one end, carries at the other end an arrangement called a *mouthpiece,* to which is fastened a reed. The clarinet reed is cut from some reasonably elastic material; that in common use is a particular variety of cane (*Arundo Donax*) resembling bamboo, which is said (by French reed makers) to grow best in France. The reed is shaped to be flat on one side. The other side is shaved down to produce a wedge shape having a thickness of about 0.1 millimeter (0.004 inches) at the thin end. The reed is fastened to the mouthpiece, clamped against a specially shaped surface called the *lay* by means of a clamping device called the *ligature.* The lay is shaped to leave a small opening between the tip of the reed and the end of the mouthpiece. The arrangement is as illustrated in Fig. 4, which shows a cross-sectional view of what we might call the basic clarinet.

To sound this arrangement, the player takes the mouthpiece and reed into his mouth and presses his lower lip against the reed to partially close the reed opening, as in Fig. 4. Air from the player's lungs is then blown through this opening into the air column of the instrument. The pressure required to do this is called the *blowing pressure.* The air flow is usually started by first covering the reed opening with the tongue,

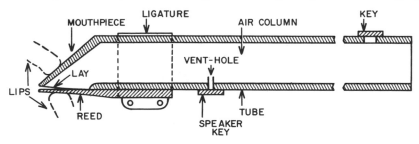

Fig. 4. Cross-sectional view of the basic clarinet.

and then pulling it away; this process is called *tonguing.* The first puff of air down the air column will set the various resonance modes of the column into vibration. The fundamental mode has a pressure antinode at the mouthpiece end. This pressure in the mouthpiece, acting on the reed, alternately pushes it farther away and pulls it closer to the mouthpiece; the size of the reed opening is thus varied at the fundamental frequency. During that half of the cycle in which the pressure in the mouthpiece is higher than the average (ambient) pressure, the opening is larger, and conversely. The air in the player's mouth is at the blowing pressure provided by the lungs, which is always higher than the highest mouthpiece pressure. If this blowing pressure is high enough, the air flow into the mouthpiece is greatest at the time during the cycle when the mouthpiece pressure is highest, because of the larger reed opening existing at this instant. This is just the condition needed for building up oscillations in the column; if air is flowing into the mouthpiece rather than out of it when its pressure is higher than average, then energy is being supplied to the system.

The behavior of this acoustical system again resembles that of the mechanical mass and spring system of Ch. 2, this time for the case where the mass is pulled a little farther out each time it momentarily comes to rest at the ends of its path. As in the case of the mechanical system, the oscillations then build up until the maximum mouthpiece pressure during the cycle is approximately equal to the blowing pressure. At this point, the energy supplied per cycle becomes small and the amplitude of the vibration remains constant at a value such that the power losses due to friction on the tube walls and to sound radiation are equal to the power supplied.

The rate of air flow into the mouthpiece depends not only on the size of the reed opening, but also on the difference between the mouthpiece pressure and the blowing pressure supplied by the player. If this blowing pressure is not large enough, air will flow out of the mouthpiece rather than into it when the mouthpiece pressure is high. The system then loses energy intsead of gaining it, and oscillations cannot build up. The practical result of this is that a certain minimum blowing pressure is required to get any sound at all, as any woodwind player knows. Considerable practice is necessary to blow sustained pianissimo tones in the woodwinds, since a momentary drop in blowing pressure below the minimum value will result in the tone stopping completely.

Since the clarinet has a pressure antinode in the mouthpiece, it resembles a tube closed at one end and open at the other. Its fundamental frequency is then half that of a tube open at both ends and of the same length. This is why the clarinet sounds approximately an octave lower than the flute, although it is of nearly the same length.

The Clarinet Scale

Having obtained the basic clarinet, next we need to make it musically usable by providing different sounding frequencies; again, we can do this by changing the length of the air column. If holes are cut in the side of the clarinet tube in the proper locations and covered with fingers or otherwise, the scale may be ascended by opening these holes one at a time, starting with the one farthest from the mouthpiece. The usual six holes would then take us up to the seventh note in the scale.

However, since the tube closed at one end vibrates with modes that are odd multiples of the fundamental, the next mode of oscillation of the clarinet arrangement has a frequency three times that of the fundamental, which musically is an octave and a fifth, or a twelfth. Therefore the clarinet will overblow the twelfth instead of the octave, and it is necessary to fill in this gap between the seventh and twelfth with additional holes and keys.

This gap in the scale is filled in the following way: the lowest note in the B♭ clarinet, using the whole length of the instrument, is written "E$_3$," sounding D$_3$ a whole tone lower. (Since many musical instruments are transposing instruments, sounding differently than written, it is convenient here to use quotation marks to indicate written notes.) The lowest note of the basic scale is "G$_3$"; by providing two additional holes and keys the two intervening semitones "F$_3$," and "F♯$_3$," are supplied. The basic scale is now ascended from "G$_3$" by opening the six holes in turn, (with additional keys provided for sharps and flats) until "F$_4$" is reached. Additional holes provided closer to the mouthpiece are then opened to get finally to "A♯$_4$." This is the highest note in the low register, also called the *chalumeau* register. The additional holes and keys required to allow the clarinet to cover the range of a twelfth in this register make clarinet fingerings somewhat more complicated than those of the other woodwinds.

To get the next higher note "B$_4$" in the scale, all the holes are again closed and the instrument is overblown to produce the next resonance mode a twelfth above the original lowest note. In the clarinet, overblowing is accomplished by opening a small hole, the *vent-hole*, in the side of the tube about 15 centimeters (6 inches) down from the mouthpiece tip. This hole essentially destroys the lowest resonance. How this happens is shown in the resonance curve for the clarinet given and discussed later on in Fig. 10. The lowest resonance moves from its position as shown by the solid line in Fig. 10 to that shown by the dotted line when the vent-hole is opened; the remaining higher resonances are essentially unchanged. With the vent-hole open, the clarinet operates in the second mode, producing the *clarinet* register. For playing in the lower register, this hole is normally closed by a cover and pad attached to a key called the *speaker key* and operated by the left thumb. The vent-hole and speaker key are shown in Fig. 4; the hole is usually provided with a short metal tube extending part way into the bore, as shown, to keep moisture condensed from the player's breath from running into the hole and clogging it.

The scale is now ascended further in the higher register by opening the holes again as before, but with the speaker key held open. The basic scale starts with "D$_5$" in this register, and goes to "C$_6$." Above this the third mode is used; the production of this mode is aided by opening one of the finger holes farther down the instrument from the speaker hole but above the other closed holes, so that there are essentially two speaker holes. A fingering such as this, which gives an open hole above closed holes, is called *cross-fingering* or *fork-fingering* and is used for many of the higher notes on the woodwind instruments. See Fig. 16(b) for an illustration of cross-fingering.

Construction of the Clarinet

The modern clarinet is constructed in five sections: the mouthpiece, barrel, top joint, bottom joint, and bell. The mouthpiece is generally made of hard rubber, plastic, or sometimes glass. The remaining sections are generally made of wood or, in some of the less expensive present-day clarinets, of plastic. The barrel is the short length connecting the mouthpiece to the top joint. The top and bottom joints carry the key mechanism and have the basic six holes, three in each section, as open holes covered by the fingers. The bell is a separate section connected to the bottom joint; it has no holes or keys. The bore of the clarinet is not cylindrical throughout its whole length; the cylindrical section begins about 4 centimeters (1.5 inches) from the tip of the mouthpiece and extends about halfway down the bottom joint with a diameter of about 1.5 centimeters (0.59 inches). From this point it flares outward to a diameter of 5.7 centimeters (2.3 inches) at the end of the bell.

As with the flutes, there exists a family of clarinets of various sizes. Most of them are transposing instruments, fingered the same for the same written note but sounding different tones according to their size. The most frequently used model is the clarinet in B♭ whose sounding range is from D_3 to about B♭$_6$, but written a whole tone higher. The clarinet in A is a semitone lower than the B♭ instrument; thus it sounds a minor third lower than written. The bass clarinet sounds an octave lower than the B♭ instrument, and the E♭ clarinet a fourth higher. The basset horn is a clarinet pitched a fourth lower than the B♭ instrument. There are also other forms, not commonly seen, which are described in the literature.[4]

Clarinet Reed Behavior

The clarinet reed, together with the mouthpiece, is a vital part of the instrument. The flat face of the reed is clamped against the flat area of the lay of the mouthpiece; the rest of the lay curves away from the reed toward the tip, as we see in Fig. 4, so that the tip of the reed is normally about a millimeter (0.04 inches) away from the tip of the mouthpiece. This is also shown in Fig. 5(a), an end view of the tip

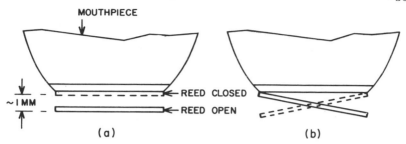

FIG. 5. End view of tip of reed and clarinet mouthpiece. (a) Normal vibration of reed tip for loud tones. (b) Unwanted type of motion producing a "squeal."

of the reed and the mouthpiece. When the instrument is played, the lower lip, together with the air pressure in the mouth, pushes the reed in to about half this distance; the reed then vibrates around this position. For soft tones the tip of the reed never touches the mouthpiece. As the blowing pressure is increased, the amplitude of the reed vibration increases until for loud tones the tip of the reed is against the tip of the mouthpiece for approximately half each cycle.[5] Such an arrangement, in which the reed touches its mounting during the cycle of the vibration, is called a *beating* reed.

The clarinet may be blown with an artificial embouchure, consisting of a box sealed to and enclosing the mouthpiece, and connected to an air supply under pressure. A piece of soft material is mounted in the box and pressed against the reed to simulate the player's lip. This arrangement allows the clarinet to sound indefinitely with an unchanging tone. (Such a tone becomes very irritating after a while.) A window is mounted in the artificial embouchure so the reed may be observed while vibrating.

With this arrangement, observation with a stroboscope (described in Ch. 8) shows that the reed is simply opening and closing like what is called a flap valve, and admitting puffs of air to the mouthpiece each time it opens. The normal motion of the reed is shown in Fig. 5(a). The tip of the reed moves up and down as shown, all parts moving with approximately the same amplitude and together in phase.

Some clarinet players are under the impression that only the corners of the reed vibrate; the reason for this belief may be demonstrated by taking the wrong end of the mouthpiece in the mouth and sucking on it instead of blowing. When this is done, the reed will be seen to vibrate at its corners. However, the vibration taking place under these conditions, as observed by a stroboscope, is as shown in Fig. 5(b) and is a rotation back and forth around the center line of the reed. This vibration is of high frequency and is one of the unpleasant squeals to which clarinets are prone; it does not take place when the instrument is playing normally.

The rotational vibration just described is not the only source of squeals; there are high-frequency modes of vibration in the air column, and the reed can excite these in the same way it can excite the fundamental, vibrating in the normal manner of Fig. 5(a). Furthermore, the reed has a resonance frequency of its own at some 2000 to 3000 cycles per second, depending on how much it is free to vibrate in the player's mouth, and it can vibrate in the normal manner at this frequency. As a result, a whole family of possible squeals exists. The damping resistance provided by the player's lip pressing against the reed is important in subduing the tendency to produce these squeals. This can be shown by trying to play the clarinet with the teeth pressing against the reed instead of the lip; the result is very unpleasant.

In its closed position, the fit of the reed against the mouthpiece is quite important. Any leakage of air into the instrument during this part of the cycle makes it harder to play, since more air is now flowing into the mouthpiece when its pressure is low rather than when its pressure is high. This is just opposite to the situation necessary to maintain air column vibrations, and means the power is being absorbed from the column instead of supplied to it. (The same thing occurs when a key pad is not tight enough and allows air to escape past it. This constitutes a *leak*, which absorbs energy from the air column vibration. The presence of a leak causes the instrument to perform poorly, and may prevent the air column from vibrating at all. The speaker hole in the clarinet is such a "leak.") If the curvature of the mouthpiece lay is not correct, the reed will not fit properly in its closed position and air can leak in along the sides; this factor is one important cause of differences in mouthpieces. A few thousandths of an inch shaved off the tip of the mouthpiece can make a considerable difference in the way it plays.

Furthermore, if the reed is warped on its tip so that it does not lie flat against the mouthpiece in the closed position, a leak will exist. Since reeds are made of a cane that is susceptible to moisture, they all warp to some extent when played; this warping will not only cause some leakage but can also affect the tone quality, as we will discuss later. The natural variation in structure of various pieces of cane makes the bending and warping of a particular reed an unpredictable matter; this means that a good reed is found more or less by chance, as every clarinet player knows. Devices exist that will measure the stiffness of a clarinet reed, but such a measurement will indicate nothing more than whether the reed will play at all; it will give no information on how well it will play.

The Oboe and Bassoon

These instruments are basically tubes with conical air columns in which the tip of the cone has been cut off and a reed attached. Since a conical air column will vibrate with both odd and even harmonics, the oboe and bassoon will overblow the octave. Both instruments use a double reed consisting of two halves of cane beating against each other. Small mouthpieces with single reeds have been tried with oboes and bassoons, but have not become popular.

The *oboe* is approximately a straight cone made of three pieces of wood: the top joint, the bottom joint, and the bell, all three sections carrying key work. The reed is attached to a conical piece of metal tube called a *staple*, which is inserted into the top joint. The top and bottom joints have in them the six holes for the basic scale, which extends from D_4 to C_5. Additional holes and keys on the lower joint and bell extend the range down to Bb_3. As in the clarinet, a *speaker hole* and key (more than one in some models) are used to play in the upper register, extending the scale from D_5 to C_6. Cross-fingerings are used to produce higher notes.

Since the fundamental frequency of the complete cone is the same as an open pipe of the same length, the oboe will have a fundamental about the same as the flute and about an octave higher than the clarinet. The instrument is also built in other sizes, the most common being the English horn, pitched a fifth lower than the oboe and carrying at the lower end the characteristic bell in the shape of a hollow sphere. Further information on these instruments may be found in the literature.[6]

The *bassoon* is a cone with total length of about 254 centimeters (100 inches). Since this is too long to play if left straight, except by someone with seven-foot arms, the instrument is folded back on itself to bring its dimensions down to manageable proportions. A sketch of the bassoon air column configuration is shown in Fig. 6(a). The conical air column is bent at approximately right angles a short distance from the small end and then bent double about halfway down. The small end of the bassoon is formed of a piece of tapered metal tubing called the *crook* or *bocal* which has a diameter of about 4 millimeters (0.16 inch) at the small end, where the reed is attached. The crook bends first upward and then down, and is inserted into the *wing joint*. This joint, made of wood, has the top three tone holes of the basic scale as normally open holes covered by the three fingers of the left hand. These holes would be too far apart to be covered by the fingers of one hand if bored straight through in the proper locations; for this reason the wood is thickened in this region, forming the "wing," and the holes

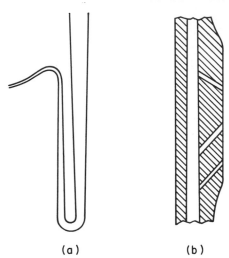

(a)   (b)

FIG. 6. (a) Cross-sectional view of the
bassoon. (b) Cross-sectional view of
the bassoon wing.

bored on a slant. This is shown in Fig. 6(b), which is a section through
the wing joint. This arrangement allows the holes to be far enough
apart along the bore and on the outside to be close enough together to
be covered by the fingers. The wing joint in turn is inserted into a section called the *boot*, also of wood, which is bored with two conical
holes along its length, one of these being the continuation of the bore
of the wing joint. The boot carries the bottom three tone holes of the
basic scale, for the right hand. A metal piece formed into a U-shaped
tube is clamped to the bottom of the boot to connect its two conical
holes so as to produce a continuous bore. Into the boot is next fitted
the *long joint*, which carries the conical bore up through the *bell*; this
is the piece carrying the characteristic metal or ivory ring seen on the
top of the instrument. At the end of the bell the diameter of the conical
bore has increased to about 4.0 centimeters (1.6 inches). The bulge on
the bell a few inches from the open end has no acoustical significance;
it strengthens the wooden wall, which would otherwise be quite thin.

The basic scale of the instrument extends from G_2 to F_3. About half
the total length of the instrument is used only to provide notes below
G_2, to extend the scale down to Bb_1. Several of the notes in this region
involve the use of the left thumb. Above F_3 the instrument overblows
the octave to get notes G_3 to D_4; this is done with the help of vent
holes, as in the other instruments. However, the large size of the instrument precludes the use of the single vent-hole as in the clarinet; the

hole must move up the instrument as the scale is ascended. This is accomplished for the lowest few notes in the higher register by partially opening the top tone hole of the basic six-hole scale (the so-called "half-holing," one of the complications of bassoon playing). For higher notes other vent-holes are provided, covered with pads and opened by keys operated by the left thumb. (Improvements in the bassoon usually involve adding another key for the left thumb, already badly overworked.) Above D_4, cross-fingerings are used to extend the scale up to C_5; higher notes than this can be played, but are quite a strain on the player's embouchure.

As with the other instruments, there are other sizes of bassoons. The only important one is the contrabassoon, which plays an octave lower than the regular bassoon. Information on the bassoon family is available in the literature.[7]

The Saxophone

The *saxophone* is a sort of hybrid; it consists of a single reed on a mouthpiece like that of the clarinet, coupled to a wide conical brass tube. It overblows the octave, like the oboe and bassoon, but otherwise has much the same fingering as the clarinet.[8] Five common sizes are available, with the lowest notes Ab_1 (bass) to Ab_3 (soprano). The soprano saxophone is straight, resembling the oboe. The other models are curved to keep the overall length to within manageable proportions; the bass saxophone has a complete loop in the crook for this reason.

Sound Radiation by the Woodwinds

In playing any of the reed instruments, the player generates a standing wave inside the instrument with a pressure antinode at the reed end and a pressure node at approximately the first open hole. The power supplied by blowing the instrument maintains the oscillations of this standing wave, as described earlier. Most of this power is used to overcome friction of the vibrating air on the sides of the tube. Sound is radiated through the open holes, most of the sound power coming out through the first open hole below the mouthpiece and most of the remainder through the second one. The actual sound power radiated by the woodwinds is only one to two percent of the total power necessary to maintain the oscillations in the instrument; like all musical instruments, they are very inefficient.[9] However, very little power is needed to produce a reasonable sound intensity, so efficiency is not important.

When a woodwind instrument is played, what one hears is of course the radiated sound, not the internal standing wave. Clarinetists are sometimes under the impression that a sound starts at the mouthpiece and travels the full length of the clarinet tube; this is not the case. Except when all the tone holes are closed, the bell does not serve as a sort of horn from which the sound is radiated. To demonstrate this, the bell of the clarinet may be removed completely, or plugged with a large cork, without making an appreciable difference in the tone of the clarinet except for the two lowest notes in each register. Similarly, the bell of the bassoon may be removed without appreciably affecting the tone of the instrument for higher notes. This is why mutes for woodwinds are not practical as they are for brass instruments, where all of the sound comes out of the bell of the instrument for all notes.

Intonation and Resonances in the Woodwinds

The woodwind instruments as actually built are only approximately the cylinders and cones on which they are based. The deviations from the simple shapes are quite important musically, since they affect both the intonation and tone quality of the instrument. To begin with, the basic six-hole, seven-note scale (together with the added holes for sharps and flats) is used in all the instruments for the two main registers, and by means of cross-fingerings for higher registers. This is possible only for air columns which have a cylindrical or conical shape. This fact can be demonstrated mathematically; it is also demonstrated from the practical standpoint in that only these shapes have survived in the evolution of the woodwind instruments.[10] For any other shape of air column, it is impossible to use the same series of holes for two different registers; if the holes are so placed that the scale is in tune in one register, it will be out of tune in the other. It follows that any deviation in the woodwind instruments from the true cylindrical or conical shape will mean that a tone hole would generally need to have two different locations to be exactly right in two registers. Therefore, in order to use the same hole for both registers, certain compromises are necessary.

The flute, for example, is different in a number of ways from a simple tube open at both ends. For all but the lowest note on the instrument, one end of the vibrating air column is actually an open side hole on the tube. The key pad for this hole is close enough to have an effect on the air column vibrations, and the remainder of the tube below the open hole also has an influence. The actual arrangement is hence more complicated acoustically than a simple open end. Likewise, the other end of the air column is the mouth hole more or less covered by the lip. In the portion of the flute between these two ends are a number of side holes covered with key pads; these add an additional volume in this region as compared to the plain tube. The length of the flute between the mouth hole and the open end with all holes closed is such as would sound C♯₄ as the fundamental; the extra volume added by the closed tone holes together with the effect of the lip covering the mouth hole brings this fundamental down a semitone to C₄.

The resonances of this rather complicated system can be measured by the external excitation method we described in Ch. 4, in which the air column is set into vibration by a sound from an external loud-speaker and the vibrations in the air column detected by means of a microphone; in the case of the flute, this microphone is installed in the cork in the head joint. When the frequencies of the first two resonance modes of the flute are so measured for the notes of the scale, it is found that they differ from the frequencies of the tempered scale by varying amounts. Furthermore, the discrepancies are not the same in the two registers, the frequencies of the modes in the upper register being somewhat more than twice those of the lower.

Fig. 7 shows the results of measurements of this kind on a particular flute.[11] The discrepancies are plotted in the vertical direction as devia-

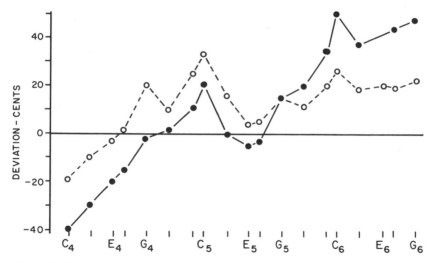

FIG. 7. Deviations in cents of the resonance frequencies and the sounding frequencies of a flute from the tempered scale frequencies, for each note in the diatonic scale. Solid curve: frequencies measured by the external excitation method. Dotted curve: sounding frequencies when blown.

tions in cents of the resonance frequencies from the values that they should have in the tempered scale. This is done for each individual note in the diatonic scale; the notes themselves are marked along the horizontal axis. The solid curve in Fig. 7 is for the resonance frequencies found by the method above; they deviate by as much as half a semitone from the tempered scale values. The upper register is sharp with respect to the lower; this is called the *octave stretch*.

When the instrument is played, the frequencies for the notes are given by the dotted curve in Fig. 7, and the discrepancies are seen to be less. The reason for this is that the player can vary the amount by which his lip covers the mouth hole, and this changes the frequency. The solid curve of Fig. 7 was made using a constant amount of equivalent mouth hole coverage. When sounding the instrument, however, the player covers more of the mouth hole as the scale is ascended, which helps to get a better tone; as a result, the tones get flatter. The taper in the head joint apparently keeps the high notes from going too flat. The frequency of a given tone can be altered from about half a semitone below to about a quarter of a semitone above the resonance frequency by changes in the blowing speed and direction, but under normal playing it is within a few cents of the resonance frequency. The design of the modern flute to bring its resonances close enough to the frequencies of the tempered scale to be musically usable is the result of much experimentation.

The clarinet has problems of a similar character. The flare of the bore has a pronounced effect on the lowest tones. The length of the B♭ instrument is 67 centimeters (26.3 inches) from the tip of the mouthpiece to the end of the bell. A simple cylindrical tube mounted on a clarinet mouthpiece with its length adjusted to give the same resonance frequency as the lowest one for the clarinet itself has an overall length of 60 centimeters (23.5 inches). The flare in the bore of the clarinet thus makes it necessary to make the instrument 7 centimeters (2.8 inches) longer to get the right lowest frequency. The reason for this is that the large diameter of the flare has a smaller acoustic mass per unit length, as we saw in the discussion of the Helmholtz resonator in Ch. 4. Hence a greater length of larger bore must be used to get an acoustic mass equivalent to that of a bore without flare. The flare in the clarinet bore has developed over the many years of empirical work on the instrument; we do not know yet the acoustical reasons for its existence.

Measurements of the clarinet resonances for different notes in the scale made by the external excitation method show that all of the resonance frequencies are above the playing frequencies.[12] This is shown in Fig. 8, where the circles show the deviations in cents of the resonance frequencies from those of the tempered scale and the vertical lines give the approximate range of the actual playing frequencies. The break between registers is shown by the vertical dotted line. The frequencies of the resonances in the higher register are not precisely three times those of the lower, again because of the effect of covered tone holes. In the clarinet a further complication is introduced by the fact

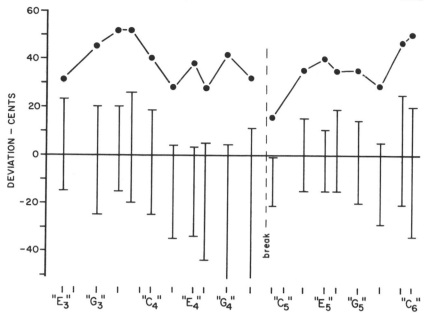

FIG. 8. Deviation in cents of the resonance frequencies and the sounding frequencies of a clarinet from the tempered scale frequencies, for several notes in the scale. Solid curve shows frequencies measured by the external excitation method. Vertical lines show ranges of sounding frequencies.

that the mouthpiece does not have a cylindrical bore out to its tip, but is wedge-shaped for part of its length at the reed end.

The playing frequencies of the clarinet are lower than the resonance frequencies because of the behavior of the reed. The tip of the reed is very thin and light and its displacement follows quite well the variations in a mouthpiece pressure. However, the damping effect of the player's lip makes it lag somewhat behind these variations of pressure, and it can be shown that this causes the actual playing frequency to be below the resonance frequency. Furthermore, the amount of this shift depends on the average gap between the reed and the mouthpiece; the smaller the gap, the higher the frequency. The player can control the gap by varying the lip pressure, so achieving the control over the playing frequency shown in Fig. 8, which is necessary to achieve correct intonation.

The oboe and bassoon have complications similar to those in the flute and clarinet. Covered tone holes are usually present. The bore of either instrument is not a simple cone; the different sections in the bassoon, for example, have conical bores of different tapers. Furthermore, the cone is not complete out to the tip, but is cut off short of the tip and the reed substituted; the bassoon would be some 30 centimeters (12 inches) longer if it were a complete cone. These factors produce deviations of the resonance frequencies from the desired tempered scale values. As in the clarinet, the player compensates for them by means of the pressure of the lips on the reed; the mechanism is the same as for the clarinet.

The bassoon demonstrates woodwind deficiencies especially well. Practically every note of this instrument is out of tune and needs to be pulled into tune with the lips; in fact, some notes can be varied as much as a minor third. This situation, together with the "half-holing" required for some notes, makes the bassoon an uncooperative and difficult instrument to play. Except for having keys added, it has changed little in several hundred years, has been aptly called a "fossil," and badly needs an acoustical working-over. (The author is a bassoonist and hopes to do this working-over if he lives long enough.)

An additional complicating factor in the intonation of woodwind instruments is the fact that their air columns are actually pure air only when they are not being played. Under normal playing they contain a mixture of air, water vapor, and carbon dioxide from the player's breath. The proportions of these constituents will vary from time to time, depending on the player's exertions, and will vary throughout the bore of the instrument, depending on how much of it is in use at a particular time. Since the speed of sound is different in these various constituents, as shown by Table I in Ch. 3, the resonance frequencies of the instrument will be continually fluctuating around their average values. The deviations are not large, but the player must compensate for them.[13]

Furthermore, since the speed of sound in a gas varies with temperature, the resonance frequencies of the instrument will change as the player's breath heats it.[14] The effect will depend on how much of the instrument is heated and the extent of the heating. It also means that the instrument will be considerably flatter when played outdoors on a cold day than when played in a heated auditorium. The frequency change with temperature, as given in Ch. 8, is 1.6 cents per degree Fahrenheit (3.0 cents per degree centigrade), so a rise in temperature from 50 to 80 degrees Fahrenheit would raise the resonance frequencies of a wind instrument by about half a semitone. This is not a negligible amount.

Tone Quality in the Woodwinds

The deviations of the woodwind bores from the simple cylindrical and conical shapes not only affect the intonation; shifting the upper resonances in the air columns also affects the tone quality. When the clarinet is blown softly, the reed does not touch the mouthpiece; its motion is nearly sinusoidal and the tone is reasonably pure. When the blowing pressure is increased to produce loud tones, the reed is in contact with the mouthpiece for a little more than half a cycle, as can be seen by stroboscopic observation. Air is then admitted into the mouthpiece in short bursts, and all the harmonics of the fundamental frequency will be present. These harmonics will cause vibrations in any modes of the air column that lie nearby in frequency.

For example, in Fig. 19, Ch. 4, there was shown a resonance curve for a tube closed at one end, obtained by the external excitation method; the resonance frequencies were quite accurately 3, 5, 7, and so on times the fundamental. If a clarinet mouthpiece with reed is placed on this metal tube and the resonances are again measured by the same method (with the reed closed against the mouthpiece), we get a curve essentially identical to that of Fig. 19, Ch. 4, except for slight changes in the positions of the high resonances (above the eleventh harmonic). These changes are due to the internal shape of the clarinet mouthpiece and are of no importance to our present discussion.

Now if the reed is blown in the usual way to produce a tone in the metal tube, an internal standing wave is obtained which contains mostly odd harmonics, since their frequencies coincide with the resonance frequencies. The even harmonics have small amplitudes, since their frequencies lie in between the resonance frequencies. In Fig. 9 is

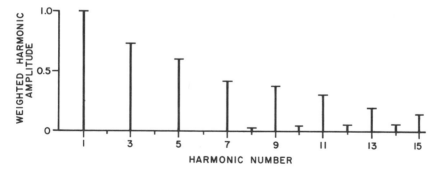

FIG. 9. Harmonic structure of the internal standing wave produced in a metal tube equipped with a clarinet mouthpiece and sounded by blowing. The actual amplitude of a harmonic has been multiplied by its number to get its weighted amplitude as plotted in the figure.

shown the harmonic structure of this standing wave. The amplitude of each harmonic has been multiplied by its number to give the *weighted* harmonic amplitude shown in Fig. 9; that is, the weighted third harmonic amplitude is three times the measured harmonic amplitude, and so on. Since it can be shown that the higher the frequency of the harmonic, the better it radiates from the instrument, this weighting gives a more realistic representation of how the ear will appreciate the actual harmonic structure. It also makes the harmonics easier to see on the graph.

Now if the same resonance measurements are made on the clarinet (again with the reed closed against the mouthpiece), it is found that they are considerably distorted as compared with those in the cylindrical metal tube, as we see in Fig. 10 for the note "E₃" on the clarinet.

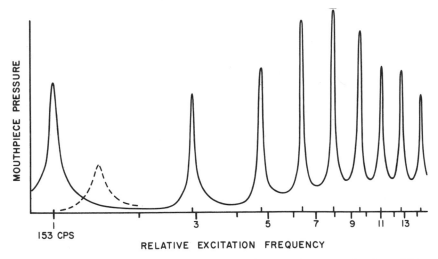

FIG. 10. Resonance curve for the note "E₃" on a clarinet. The dotted line shows the position of the lowest resonance with the vent-hole open.

The resonance frequencies depart more and more from the harmonic frequencies as we go to higher values; for this particular note, the eighth harmonic, for instance, coincides more nearly with a resonance than either the seventh or ninth. Now if the clarinet is blown and the harmonic structure of the internal standing wave is analyzed, we find

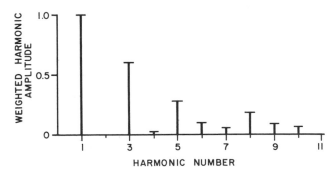

FIG. 11. Weighted harmonic structure of the internal standing wave in a clarinet sounding "E₃."

that the eighth harmonic is more prominent in this tone than the seventh or ninth, as might be expected since it lies close to a resonance; this is shown in Fig. 11.

As the scale is ascended in the clarinet the resonances become fewer; for the note "A$\sharp_4$" just before the break to the higher register, there are only two good resonances, as shown in Fig. 12. As a consequence, the tone has few strong harmonics, as we see in Fig. 13. The lack of reso-

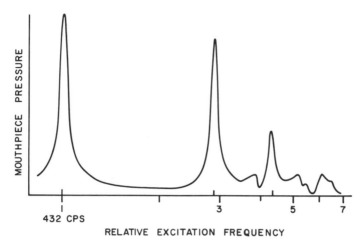

FIG. 12. Resonance curve for the note "A$\sharp_4$" on a clarinet.

nances in this region accounts for the poor quality of the so-called *throat tones* in the clarinet.

In the higher register, above the break, there is little correspondence between the harmonic frequencies and the resonances, as may be seen in Fig. 14. The third harmonic is the only one near a resonance, and it is not very close. As a consequence, the tone produced is mostly funda-

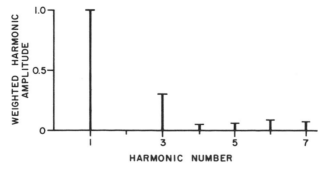

FIG. 13. Weighted harmonic structure of the internal standing wave in a clarinet sounding "A$\sharp_4$."

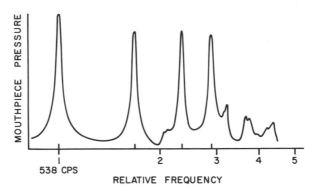

FIG. 14. Resonance curve for the note "D_5" on a clarinet.

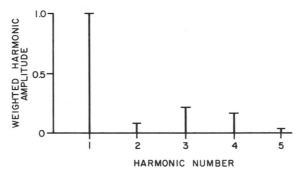

FIG. 15. Weighted harmonic structure of the internal standing wave in a clarinet sounding "D_5."

mental but has both even and odd harmonics, as shown in Fig. 15. For this reason the quality of the clarinet tone is different in the higher register.

Measurements of resonances in the bassoon show the same general picture. For the low tones of the instrument, several resonances appear, lying reasonably close to the harmonic frequencies. As the scale is ascended, the resonances become fewer in number and depart more from the harmonic frequencies. Above the break in the bassoon there is again little correspondence between the resonances and the harmonic frequencies.

In the bassoon, that portion of the instrument below the first open tone hole has a considerable influence on the resonances of the portion of the instrument producing the tone, much more so than in the other woodwinds. Closing or opening a tone hole in the lower part of the bassoon can make quite a change in the quality of a tone higher up on the instrument; for this reason bassoonists have a number of fingerings available to improve the quality of certain bassoon tones.

Much more work needs to be done on the bassoon as well as the other woodwinds to work out the relationships between the structure of the instrument, the positions of the resonances, and the harmonic structure of the tone. At present we are not only ignorant of those relationships, but do not even know in physical terms the difference between a "good" tone and a "bad" tone.

In the early days of woodwind instruments, cross-fingerings were used for the chromatic tones. If we sound a tone on the instrument (not using its full length) and then close the hole just below the first open hole, leaving the remaining holes open, the pitch of the tone drops approximately a semitone. However, the tones produced in this way are of poor quality, and the use of cross-fingerings has become obsolete; instead, the chromatic tones are supplied by additional tone holes provided with pads and keys. The reason for the poor quality of cross-fingered tones is the effect on the higher modes of the instrument of the portion of the tube below the first open hole. Cross-fingering puts a longer section of tube below the first open hole, as shown in Fig. 16.

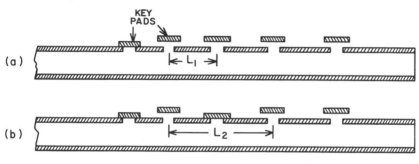

FIG. 16. (a) Normal fingering. (b) Cross-fingering.

For normal fingering, a length L_1 of tubing is below the first open hole; for cross-fingering, a greater length L_2 is below the first open hole. These are shown in Fig. 16(a) and (b). The effect of this greater length is to push the higher modes of the air column even farther away from the harmonic frequencies than they are normally, so the harmonic content of the tone is further reduced; this results in a poorer quality.

Other Factors in Woodwind Tone

Woodwind instrument reeds have a very pronounced effect on the quality of the tone produced by the instrument, as any player knows. As we pointed out, a clarinet reed, in its open position away from the mouthpiece, can vibrate at a high frequency whose value depends on the amount of reed extending past the player's lip, and which lies somewhere above about 2000 cycles per second. This will tend to increase the amplitude of the harmonics whose frequencies lie near that of the reed. Furthermore, in its closed position, a slight warping of the end of the reed can leave portions of the reed free to vibrate because they are not in contact with the mouthpiece. High-frequency vibrations in these portions of the reed can then be produced during that part of the cycle. Oboe and bassoon reeds can behave similarly. As a consequence, the woodwind reed is a rather complicated and unpredictable structure, and is a continual source of trouble. A great deal more investigation is needed before we understand reed behavior. Synthetic materials have been tried for reeds, but so far have not proven successful, since the details of the qualities necessary for a good reed have not been known. With all of the materials now available from modern technology, it should eventually be possible to make a good synthetic reed with uniform and predictable qualities.

The material from which a woodwind instrument is made has long been considered an important factor in the quality of its tone.[15] Arguments about the best material for wind instruments probably started in early Stone Age circles with assertions that a flute made from a human thighbone had a much better tone than one made from a stick of bamboo. The term "woodwind," of course, comes from the fact that originally such instruments were all made of wood; however, at the present time flutes are more often made of metal. There seems to be a feeling that the more expensive the metal, the better the flute; gold and even platinum have been used. On the other hand, metal clarinets have never been popular and most players disparage them, notwithstanding tests showing that good wood and good metal clarinets have practically identical tones.[16] Books on woodwinds speak with authority on the tones produced by various materials, but without any justification for such statements.[17]

The belief in the importance of material to woodwind tone probably arises from the observation that when the instrument is played, there are sometimes vibrations in its body that can be felt by the fingers. (In the clarinet, these vibrations are mostly those communicated directly to the body by the reed.) It is then assumed that the vibrations of the walls can radiate into the air sound that is part of the tone, and that they can also affect the internal standing wave and change its quality. However, recent work has shown that the sound radiated from the walls is negligible compared with the sound normally radiated by the instrument, and that the effect of the wall vibrations on the internal standing wave is also negligible.[18] It follows that the material of the instrument is not a factor of its tone; if a gold flute sounds different from a silver one, it is because of differences in the structural dimensions and not because of the differences in the metal. A piece of flexible plastic tubing provided with a clarinet mouthpiece and bored with a few tone holes in the right places will sound enough like a clarinet to fool people who do not see what is being played. In short, the material of the wind instruments may be chosen for its working qualities and not for any imagined effect on its tone.

The only part of the woodwind reed that vibrates appreciably is the end inside the player's mouth. In the clarinet, for example, that portion of the reed outside the lips does not vibrate any more than does the body of the instrument to which it is attached. It follows that various schemes involving the cutting of grooves, holes, and so forth in the thick part of the clarinet reed to supposedly improve its tone are useless. This also applies to ligatures made with the tightening screws on top of the mouthpiece instead of underneath.

The mouth and chest cavities of the player are also sometimes considered to have an effect on the tone the instrument produces. This may be a factor in the case of the flute, but at least in the case of the clarinet such work as has been done indicates that this factor is of no importance. However, this point needs further investigation.

The Pipe Organ

The *pipe organ* is not usually classified as a woodwind instrument, but since it uses the same sound-producing mechanisms we may consider it briefly here. It is an assemblage of some hundreds of resonant tubes of various shapes, some, called *flue pipes,* using the edge-tone mechanism to produce sound, and others, called *reed pipes,* using a vibrating reed.

Each pipe is mounted on an air chest supplied with air under a pressure from a blower. The pressures used are measured in *inches of water*—i.e. the height in inches that a column of water in a U-shaped tube will be raised when the given pressure is applied to one side. Pressures commonly used are in the range of three to eight inches of water. A valve in the chest under a particular pipe may be opened by pressing a key on the organ manual, admitting air into the pipe. The amount of air is regulated by the size of the hole in the foot of the pipe.

In the flue pipe, shown in Fig. 17(a), the air flows through a slit between the lower lip of the pipe and a cross barrier called the *languid.* The stream of air from this slit strikes the upper lip and sets the air column of the pipe into vibration in the manner already described for the edge-tone mechanism. The vibration frequency is of course determined by the length of the pipe and by whether it is open or closed at the end away from the mouth. Pipes are tuned by changing their overall length. This is done by means of a moveable plunger in the closed pipe, and by means of a sliding collar or by rolling down a strip of metal formed by lengthwise cuts in the open pipe, as shown in Fig. 17(a).

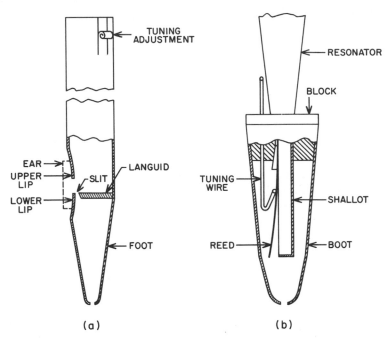

FIG. 17. Organ pipes. (a) Diapason pipe. (b) Reed pipe.

The organ manual commonly consists of 61 keys, extending from C_2 to C_7. An open pipe giving the frequency of 65.4 cycles per second corresponding to C_2 will have a length of about 8 feet, as given by Eq. (8), Ch. 4. The series of pipes covering the range C_2 to C_7 is thus called an 8-foot *rank;* the same designation would be used for any series of pipes giving the same frequencies. Correspondingly, a series of pipes coupled to the organ manual but sounding an octave lower would be designated as a 16-foot rank. Thus a series of closed pipes starting with one 8 feet long would be a 16-foot rank.

Various shapes of pipes give various tone qualities. Closed pipes have mostly odd harmonics in the tone, whereas open pipes have both odd and even harmonics. The commonest pipe is the so-called *diapason,* which is a cylindrical open metal pipe. The resonances of the pipe and hence the harmonic structure of its tone depend on the ratio of the diameter of the pipe to its length. If this is small, the influence of the mouth and open end are small, so the resonances correspond rather closely to the harmonic frequencies. The tone of the pipe is then rich in harmonics. Conversely, if the pipe is of large diameter as compared to its length, the higher resonances get rather far out of tune with the harmonics, so the tone is mostly fundamental. The use of shapes other than cylindrical also provides for different tone qualities.

The harmonic structure of the tone of the pipe is affected greatly by the process of *voicing*, which consists of adjusting the width of the slit, the position of the languid, the position of the lip, and so forth. Its purpose is to get the pipes in a given rank to sound alike and with the proper tone. The pipe voicer works on the basis of previous experience, there being at present no scientific foundation on which to base the process.[19]

The reed pipes in the pipe organ use metal reeds placed at the bottom of a conical tube, as shown in Fig. 17(b). The reeds used may be beating reeds, as described for the clarinet, or free reeds, in which the reed swings back and forth through a slot in its support without touching it. In either case, the resonance frequency of the system depends both on the frequency of the reed and on the frequency of the attached tube. It can be changed over a sufficient range by altering the vibrating length of the reed by means of a wire which presses against the reed and can be adjusted from outside the pipe. The metal cone above the reed reinforces certain harmonics of the sound produced by the reed, and adds its character to the tone.

As in the case of the wind instruments, the material from which organ pipes are made has long been considered to exert a strong influence on the tone of the pipe.[20] Tin especially has been considered an important constituent.[21] Pure tin is assumed to make the best pipes, but so-called "spotted metal," a mixture of tin and lead in varying proportions, is generally used to save on cost, since tin is an expensive metal. For large pipes, even spotted metal is generally not used, but instead is replaced by pure zinc. The effect of the material on the quality of the pipe tone has been argued about pro and con for years.[22] However, it now appears that the material is of no significance to tone. An organ pipe made of wrapping paper can be made to sound as good as one made of tin.[23] As in the woodwinds, recent work has further shown that the vibration of the walls of the organ pipe when sounding do not affect the internal standing wave, nor do they radiate sufficient sound to be heard.[24] In practice, the most compelling reason for the use of the usual tin-lead mixture is the ease with which it may be worked and the pipe voiced.

Further information on the pipe organ may be found in the literature.[25]

The Voice

The human voice combines the basic principles of the other instrument families in a unique way. It is fundamentally a wind instrument, since it is operated by air under pressure, but it does not depend on well-defined resonances to determine frequencies, as the woodwinds and brasses do. Instead, it operates much like a string instrument; vibrations are produced by pieces of tissue whose vibration frequency, as in the strings, can be adjusted by changing their tension. The tones these vibrations produce have many harmonics. The relative amplitudes of these harmonics are altered by the resonances in the system, which thus determines the quality of the tone that finally emerges. The distinguishing feature of the human voice is the fact that these resonances are instantly adjustable, so that the harmonic structure of the voice tone can be changed quickly over a considerable range.

A sketch of the vocal tract is shown in Fig. 18. Air under pressure from the lungs is forced through the windpipe or *trachea*, which is

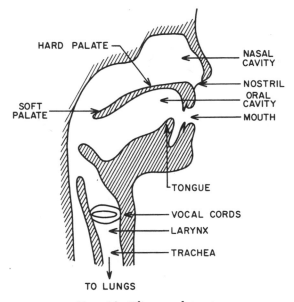

FIG. 18. The vocal tract.

terminated at its top by the sound-producing structure called the *larynx*. This contains the *vocal cords,* which are set into vibration by air flowing through them from the lungs. The vibration frequency of the vocal cords is varied by changing their tension by means of attached muscles. Above the larynx is the *oral cavity*—the space inside the mouth, which can be connected to the outside by opening the mouth—and the *nasal cavity*—the space behind the nose, which is connected to the outside through the nostrils.

The space between the vocal cords is called the *glottis*. Motion picture photographs of vibrating vocal cords show that the glottis opens up and closes down once each cycle and usually closes completely during part of the cycle. The air flow through the glottis is then in the form of discontinuous puffs, much like the air flow through the reed opening into the clarinet when it is blown. The tone the vocal cords produce is thus composed of a considerable number of harmonics.

The nasal and oral cavities above the larynx resemble Helmholtz resonators and so have resonance frequencies that depend on the volumes of the cavities and the areas of the openings to the outside. That for the nasal cavity is fixed, but that for the oral cavity can be changed by varying the volume of the cavity with the tongue and by changing the area of the mouth opening with the lips. Those harmonics of the tone from the vocal cords that have frequencies near the resonances of these cavities will be reinforced. The resonances of the nasal and oral cavities thus provide formants in the vocal tract.

We discussed formants earlier in Ch. 6 in connection with the quality of musical sounds. Whether they are important in the behavior of musical instruments is a point not yet settled, but there is no doubt that they determine voice quality. The subject of voice quality is of considerable importance to the communications industry, which has done a great deal of research on speech sounds. As a consequence, it has been found that those voice qualities that we associate with vowel sounds are determined by the formant frequencies. For example, the vowel ē as in "eat" has formants in the neighborhood of 300 and 2300 cycles per second; the vowel ā as in "ate" has formants around 500 and 1900 cycles per second.[26]

The way the formants reinforce harmonics in the tone of the vocal cords is illustrated in Fig. 19. The higher formant of the vowel ē reinforces harmonics in its neighborhood, as is evident from Fig. 19. The harmonic structures of tones of several different fundamental frequencies are shown, and in each case a group of harmonics in the region around 2300 cycles per second shows up strongly.

The importance of formants to voice quality can be demonstrated by an amusing experiment. The speed of sound in helium gas is about 2.7 times that in air, as we may see from Table I, Ch. 3. Helium is an inert gas, and the lungs may be filled with it for a short time without harm. If this is done, the oral and nasal cavities will also be filled with helium, and their resonance frequencies will go up more than an octave. This displaces the formant regions upward by this amount. The frequencies of the vocal cords, on the other hand, are not affected. Spoken sounds under these conditions will have a startling and amusing "Donald Duck" quality that shows in a striking manner the effect of formants on speech sounds.

In contrast to the extensive work done on speech sounds, there has been very little research done on the singing voice.[27] Measurements of the power output of the voice have been made; the maximum is about one watt.[28] We know practically nothing about the qualities necessary

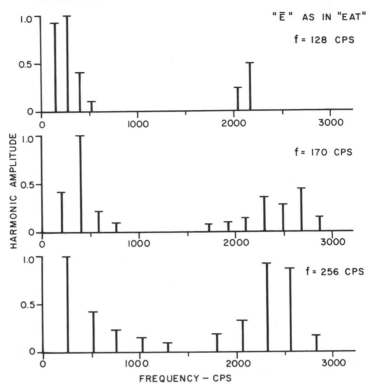

FIG. 19. Formant in the vowel sound ē as in "eat."

to a good voice. Some work has been done on the voice vibrato. Like that on the violin, the vibrato is a periodic change in frequency occurring some six to seven times per second, accompanied usually by an amplitude variation at the same rate.[29] The frequency variation is greater than one would expect, amounting sometimes to as much as a semitone; however, the ear averages out this variation and fixes the pitch as halfway between the two extremes.

To repeat, we know practically nothing about the qualities necessary to a good voice. It seems generally agreed that a good even vibrato—one whose amplitude and frequency are constant—is essential to a good voice.[30] This is about all we can say at the present state of our knowledge of the acoustics of singing. It has been claimed that specific formants are necessary for good voice quality, but this has been argued.[31] It has also been suggested that the sub-glottal region—the part of the vocal tract below the vocal cords—is important, but this is not established. In short, the whole field of the quality of the singing voice needs a great deal more research.

CHAPTER TWELVE

THE BRASS
INSTRUMENTS

The brass instruments, like the woodwinds, have their origin in pre-historic times. Primitive man found that if holes of the proper shape were provided on seashells, hollow animal horns, and the like, and these holes were blown with the lips in a certain fashion, there would result a loud and satisfying blast of sound.[1] Archaeologists have found specimens of shells provided with such holes for this purpose. Origi-nally, their most likely use was to frighten enemies, rather than for any musical purpose; however, from these humble beginnings there evolved the modern brass instruments. Originally the primitive musician used whatever natural shapes were at hand; later, with the development of techniques and skills of metal working, the simpler natural shapes could be copied in copper, brass, or whatever materials were suitable. In the course of time these metal counterparts would be altered and improved according to musical requirements, and so they eventually acquired the shapes they have today.

Acoustically, the brass instruments might be classified with the wood-winds as wind instruments; however, they differ in enough important respects from the woodwinds to merit separate discussion. First, the vibrations of the air column in the brass instruments are maintained by the vibrations of the player's lips, instead of by air streams or reeds. Since the lips are considerably more massive than woodwind reeds, they can more easily influence the air column vibrations. Second, the brass instruments use the various resonance modes of the air column, as do the woodwinds, but use many more of them. Some brass instruments use only these resonance modes; the majority, however, have a complete scale, made available by filling in the gaps between modes.

Third, to obtain the notes between the resonance modes, the brass instruments increase the overall length of the air column by inserting additional pieces of tubing, rather than by opening holes bored in the side of the tube to reduce the length. Since there are no open holes in the walls of the tube, all the sound of the brass instrument must come out of the bell. These factors make the behavior of the brass instruments quite different from that of the woodwinds.

A fourth difference, not as obvious, appears when we compare the acoustical foundations for the two groups of instruments. In Ch. 11 we saw that the woodwind instruments utilize the well-defined and harmonically related resonance frequencies of cylindrical tubes and cones. The presence of covered tone holes and other factors produce unavoidable deviations of the bores of actual instruments from these simple shapes. These deviations shift the resonance frequencies away from the harmonic frequencies, with resultant intonation problems. In the brass instruments, on the other hand, distortions from the simple shapes are introduced deliberately and in large degree in order to produce a musically useful series of resonance modes. Let us consider in some detail how this is done.

The Acoustical Evolution of the Brass Instruments

If the lips are placed on a smooth ring and put under some tension, they can be made to vibrate by blowing air through them from the lungs, producing a buzzing sound. The "Bronx cheer" is an insulting illustration. Since the mass of the lips is relatively large, the vibration frequency they produce is relatively low. If the tension of the lips is increased, they will vibrate at a higher frequency, just as will vibrating strings.[2]

If the ring against which the lips are pressed is now attached to a tube, the vibrating lips will build up oscillations of that particular resonance mode of the air column which has a frequency near that of the lips. As these oscillations increase in amplitude, the pressure they create in the air column will react back on the lips. At that portion of a cycle where the pressure just outside the lips is higher than the average pressure, it will help the lips to open. This allows more air to flow into the air column from the mouth, since the blowing pressure inside the mouth is maintained higher than the highest pressure in the air column. Energy is then supplied to the air column during this part of the cycle, and the air column vibrations will build up to a constant amplitude. The mechanism for producing vibrations in this system is then basically the same as that for the reed instruments and, similarly, there will be a pressure antinode in the air column at the lip end, where energy is being supplied.

The frequency of the combination of lip and air column will be primarily determined by the frequency of the particular mode excited. However, again as in the reed instruments, the lip can influence the vibration frequency, and because of the relatively large mass this influence is considerable, more so than for the woodwinds. For this reason, increasing the lip tension can fairly easily excite higher modes of vibration of the air column; the vent-holes needed on the woodwinds are not necessary. This makes it possible to use many more of the resonance modes of the air column than do the woodwinds.

Since there is a pressure antinode at the lip end of the column, this end behaves as if it were closed. We should then expect that if we sound a tube by blowing it in this manner, we should get modes of vibration whose frequencies are 1, 3, 5, and so forth times the fundamental frequency, as in the clarinet. This is easy enough to check experimentally. We take a piece of tubing of 1.2 centimeters (about 0.5 inches) inside diameter and some 127 centimeters (50 inches) long (somewhat shorter than air column length of a B♭ trumpet), and put on one end a ring (about like that on a trumpet mouthpiece) to make the lips comfortable. By blowing on this arrangement, we can get a series of notes that sound something like a badly-out-of-tune bugle call; the notes obtained with the length given are approximately as shown in the staff in Fig. 1. The actual frequencies as measured in the labora-

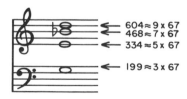

FIG. 1. Notes blown by the lips on a piece of tubing.

tory are as given, and a little arithmetic shows that they are 3, 5, 7, and 9 times a value, as shown, which we may assume is the fundamental. This fundamental is too low to be blown with the lips; however, it can be sounded by putting a clarinet mouthpiece on the tube and blowing it. A little arithmetic with Eq. (9), Ch. 4, shows that this fundamental has the right frequency for this length of tubing.

The resonance frequencies of the simple tube may be measured as was done for the clarinet by closing off the lip end, inserting a microphone at this end, and using the external excitation method we described in Ch. 4. When this is done, we find that the resonance frequencies of the tube coincide very nearly with the frequencies at which it blows. This is true of the brass instruments in general.[3] Since this is the case, we can use the external excitation method to determine the resonance frequencies of the brass instruments as well as the simple tube, and so learn something about their behavior. This will have the advantage of avoiding complications due to the influence of the lips; it will also show the fundamental, which could not be obtained by blowing with the lips.

The actual numbers representing the resonance frequencies are not too meaningful by themselves; we can understand them better by plotting them on a semitone scale. This is done in Fig. 2, where a scale of semitones is marked off by horizontal lines. For convenience, certain notes are identified along the right-hand side of the figure. Any frequency may now be converted to semitones and cents and plotted on this scale. Furthermore, to make the figure more intelligible to the musician, a somewhat distorted musical staff may be superimposed on the semitone scale. The reason for the distortion is that there are four semitones in the interval G_4–B_4, and only three in the interval E_4–G_4, and similarly elsewhere.

The resonance frequencies, including the fundamental, as obtained for the simple cylindrical tube closed at the lip end, are plotted in the left-hand column of Fig. 2. It is immediately obvious that these frequencies are far from those we associate with the usual sounding modes of a B♭ trumpet. Since we are dealing with sounding frequencies rather than written notes, we would expect to get B♭$_3$, F_4, B♭$_4$, D_5, F_5, and so on, for the trumpet; instead we see for our tube that the modes are too stretched out, being too low for the first three and too high for the remaining ones. Our simple tube does not have a musically useful series of modes.

From the musical standpoint, the tube has another fault not yet mentioned. The tone it produces when blown with the lips is subdued, muffled, and of poor quality, nothing like what we associate with a good brass tone. Hence the simple tube needs a major overhaul to make it into a good musical instrument. To a reasonable extent the makers of brass instruments have accomplished this overhaul. This has been done for the most part in complete ignorance of the acoustical basis for it, an ignorance which persists even to the present. However, as a result of many years of cut-and-try, empirical methods have been worked out for shifting the modes of the simple tube to musically useful places and at the same time improving its sound output and tone. This is done by two major alterations: a mouthpiece is added to the lip end of the tube

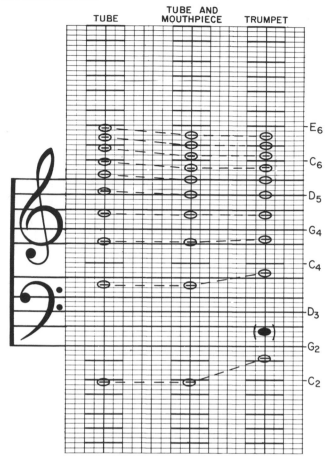

Fig. 2. The evolution of a trumpet. Left: resonances in
a simple closed tube. Center: resonances in a tube with a
mouthpiece. Right: resonances in a trumpet.

and a properly shaped flare called the *bell* is formed at the other end.
The shape of both is vitally important.

Effect of the Mouthpiece

The mouthpiece consists of a small *cup*, with a rim to accommodate the
lips. This cup connects to a tapering tube of considerably smaller diameter than the rest of the instrument; this small tube is called the *back
bore*. A section through a trumpet mouthpiece is shown in Fig. 3(a).

If this mouthpiece is attached to the tube we have been considering
and the cup is assumed closed off by the lips, the tube will be effectively

FIG. 3. (a) Trumpet mouthpiece. (b) Horn mouthpiece.

lengthened. At low frequencies, where the sound wavelengths are long compared to the size of the mouthpiece, it turns out that the amount of lengthening is just that of a piece of the tube having the same volume as the mouthpiece. For example, a given mouthpiece was measured and found to have a volume of 6.5 cubic centimeters. This is the same volume as a length of 5.7 centimeters of our tube (of 1.2 centimeters inside diameter). Hence if we cut 5.7 centimeters off the tube and attach the mouthpiece to it, the lowest resonance frequency will be unchanged.

The higher resonances, however, will not remain unchanged when the mouthpiece is added. The combination of cup (when closed by the lips) and tube is essentially a Helmholtz resonator, as described in Ch. 4, and so has a certain resonance frequency. At its resonance frequency it behaves the same as a tube closed at one end that has this as its fundamental frequency. For the mouthpiece above, for example, the resonance frequency measured 639 cycles per second, the same as that for a closed tube of length 13.5 centimeters. Hence at this frequency the mouthpiece effectively adds this much to the length of the tube. The overall result is that the mouthpiece makes the tube look longer by an amount that increases as the frequency increases.[4] As a consequence, our shortened tube with the mouthpiece attached has the same lowest resonance frequency as before, but the high modes are shifted down from their original values. The result is shown in the center column of Fig. 2. The mouthpiece has pushed the high modes down to where they belong, from D_5 on up, as indicated by the dotted lines in the figure.

The Effect of the Bell

The next step is to move the lower modes up. This is accomplished by flaring the open end of the tube, increasing its diameter as the open end is approached, and so forming the familiar bell characteristic of the brass instruments. The addition of the bell raises the frequencies of the lower resonances, the lowest being shifted up the most. If at the same time the total length is properly adjusted, the high resonances will be left unchanged. The actual amounts that the lower modes are shifted will depend on the detailed shape of the bell. With a properly shaped bell and mouthpiece, the resonance frequencies will be shifted into those modes we customarily think of as belonging to the trumpet.

In Fig. 2(c) are plotted the resonances for an actual trumpet. These were obtained from the resonance curve of the instrument made by the external excitation method; the resonance curve itself is shown in Fig. 4. The second, third, and higher modes are now in the proper places, forming approximately a harmonic series; their frequencies are very nearly 2, 3, 4, and so on times a fundamental frequency. This fundamental frequency is indicated by the black oval in parentheses in the right-hand column of Fig. 2. However, the rather startling fact evident in the figure is that this fundamental does not exist as an actual resonance mode in the trumpet. We will therefore refer to it as the *fictitious fundamental*. The lowest actual resonance mode in the trumpet, which has been shifted up by the bell (but not far enough) is more than a fourth below the fictitious fundamental, customarily thought to be the lowest resonance.

It is a common fallacy to think that a trumpet is like a tube open at both ends, giving the fundamental and higher harmonics, and that the fundamental is not used because it gives a poor tone. We see instead that the actual fundamental in the trumpet—that is, the lowest resonance—is very much out of tune with the other modes; in practice, this is immaterial, since it is not used. The so-called *pedal tone* in the trumpet—the octave below the first used mode, and hence the fictitious fundamental—does not exist as a resonance, but is produced by buzzing the lips at the right frequency. The production of the pedal tone is aided by the fact that its harmonics very nearly coincide with the existing higher modes in the instrument, even though the fundamental itself is nowhere near a resonance mode.[5] In other words, the presence of harmonics 2, 3, 4, and so on in the tone will cause the air column vibration (and hence the pressure in the mouthpiece) to repeat at the fundamental frequency. This will help the lips to vibrate at this frequency even if the fundamental itself is not present.

Adding the bell to the trumpet has another very important effect; the tone it produces is much louder and clearer than that produced by the cylindrical tube without the bell. There are two reasons for this. First, the amount of sound which can radiate from the open end of a vibrating air column depends on the area of the open end, as one might expect. The bell of a brass instrument serves the very important function of increasing the sound output of the instrument. (This may be demonstrated by attaching an ordinary tin funnel to the simple tube; its loudness is considerably increased.) Second, the production of harmonics in the tone is helped. Suppose we sound G_3 in the simple tube; its harmonics will be approximately G_4, D_5, G_5, B_5, D_6, and so on. From

Fig. 2(a), however. we see that except for D_5, none of these lies very close to a resonance, so their production will be poor. For the trumpet, on the other hand, the harmonics of Bb_3, for example, all very nearly coincide with resonances, so they will all be produced in good measure and give the tone its characteristic brass quality.

The resonance curve for a trumpet, as obtained by the external excitation method, is shown in Fig. 4. For this curve a relative frequency

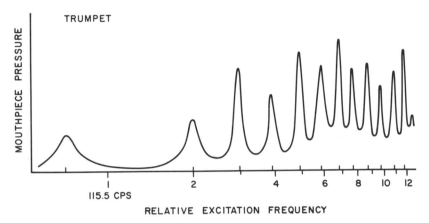

FIG. 4. Resonance curve for a trumpet, made by the external excitation method.

scale has been drawn along the bottom in terms of the fictitious fundamental of 115.5 cycles per second, and the resonance curve has been plotted so that the second resonance mode coincides with relative frequency 2, of 231 cycles per second. The positions of the resonance modes are shown by short vertical lines above the horizontal axis. Those above the second resonance very nearly coincide with the relative frequencies 3, 4, 5, and so on, and so form an approximate harmonics series. For the trumpet used, there were a few small resonances past the 12th; these are not shown in Fig. 4.

It is sometimes asserted that the brass instruments play in just intonation. This is a mistaken conclusion that comes from the application of the term "harmonic series" to the resonance modes of the brass instruments. As we see now, the harmonic relationships are not necessarily exact. Since the modes are put into their final positions by shaping the bell, a shape that is not quite right will result in discrepancies in the positions of the modes;[6] actually, there is no reason for assuming that there is any shape that will put all the modes into exact harmonic relationship. As an illustration, the resonances for a particular trumpet were measured by the external excitation method. If the instrument were tuned to give resonances whose frequencies are exactly in just intonation, their values would be precisely 2, 3, 4, and so on times the fictitious fundamental. Instead, the measurements gave results as follows: if the second resonance was taken as exactly twice the fictitious fundamental, the others were flat by the following amounts:

Third resonance   —   6 cents
Fourth resonance   —   25 cents
Fifth resonance   —   34 cents
Sixth resonance   —   16 cents
Eighth resonance   —   54 cents

The octave "C_4"–"C_5" is then 25 cents flat; "C_5"–"C_6" is 29 cents flat, and similarly for other intervals. These discrepancies will presumably have to be compensated for by the player. Hence, it is not correct to think that the trumpet, or any other brass instrument, automatically plays in just intonation.

Filling the Intervals Between Modes

The instrument as developed to this stage, having modes that fit a useful musical sequence, has its practical application in the *bugle,* with its characteristic calls played by permuting the notes of the approximate harmonic series. However, the bugle is rather limited musically, so the next step is to fill in the gaps between modes to provide the instrument with a complete scale. In the woodwinds this is done essentially by shortening the tube by opening holes bored along its length in the tube wall. The clarinet, for example, requires some eighteen holes to fill in the interval of a twelfth between the first two resonance modes. In the brass instruments the job is somewhat easier; the largest interval is the fifth, between the first two usable modes, requiring six additional semitones to fill the space. These are provided for by lengthening the air column. Six additional lengths are sufficient; in addition to filling the space between the first two modes, they allow the instrument to play six semitones below the lowest usable mode and are more than adequate to fill in the gaps between higher modes.

The simplest way to lengthen the instrument is illustrated by the system used for the *trombone,* in which a U-shaped piece of tubing is moved back and forth to shorten and lengthen the air column. This method is possible because the tubes are cylindrical; it would obviously be impossible with tapered tubes. In the trombone the familiar slide arrangement provides six additional lengths to give the additional six

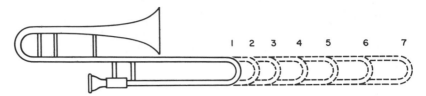

FIG. 5. Playing positions of the trombone.

semitones. The positions of the slide are shown in Fig. 5, which shows the seven positions the trombone player must learn if he is to stay in tune with the other instruments. Going down a semitone amounts to a 6 percent reduction in frequency, corresponding to about a 6 percent increase in length. Since the length of the instrument increases as the scale is descended, each additional length is slightly more than the previous one, as Fig. 5 demonstrates. The trombone player learns this somewhat warped series of positions by thousands of repetitions.

The trombone is the only member of the brass family using this simple system. In the remaining instruments the lengthening process is accomplished by inserting short lengths of additional tubing by means of an arrangement of devices known as *valves*. The mechanism of a single trumpet valve is shown in Fig. 6. A cylindrical piece of metal called the *piston* is arranged to move up and down inside a piece of tubing soldered into the cylindrical portion of the trumpet tube, as shown. The piston is normally held in the up position by a spring, but can be pushed down with the finger. In the up position, a hole in the piston called the *windway* carries the air column straight through the valve, as we see in Fig. 6(a). In the down position, two other windways in the piston divert the air column as in Fig. 6(b) so it passes through an additional piece of tubing. Pushing the valve down thus lengthens the air column; the amount of increase depends on the length of the added valve tubing. This is made adjustable for tuning purposes by providing it with a moveable U-shaped portion called the *valve slide,* as we see in Fig. 6(b).

It is very important that the pistons in the valves fit closely enough inside their cylinders that there can be no leakage of air past them. Any leaks at the pistons will have the same unfortunate effect on the resonances as do leaks in the woodwinds, making the instrument stubborn and uncooperative.

In the trumpet, three valves are used. In the usual arrangement, the second valve (counting from the lip end) adds tubing which shifts the frequency down one semitone. The first valve shifts it down two semitones, or a whole tone. The combination of values 1 and 2 then gives

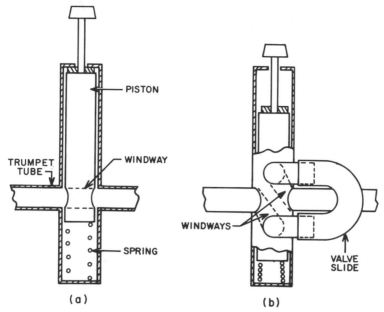

FIG. 6. Structure of a trumpet valve. (a) Piston up. (b) Piston down.

a shift of three semitones, or a minor third. The third valve alone also gives a shift of approximately three semitones, so valves 2 and 3 give four semitones, valves 1 and 3 five semitones, and all three valves six semitones.

There is a fundamental difficulty with this arrangement of adding tube lengths by combining valves, which shows up immediately. If the first and second valves each insert pieces of tubing of the correct length to lower the pitch by a whole tone and a semitone respectively, the two valves together will add tubing too short to lower the pitch by three semitones. We may see this from the slide position of the trombone in Fig. 5. Since each successive increment of length is longer than the preceding one, the amount of additional tubing needed to go one semitone down from third position to the fourth position is more than that actually added in going down one semitone from the first to the second position. In other words, the first valve must add a length of tubing equivalent to going from position 1 to position 2 in the trombone, and also from position 3 to position 4; unfortunately these are two different lengths.

We may illustrate the problem further by working out some actual numbers. The trumpet is about 137 centimeters (54 inches) long; its equivalent length—that of an open tube without bell or mouthpiece to give the same lowest usable mode, B♭₃—is about 148 centimeters (58.3 inches). To lower the pitch by a semitone requires dividing the frequency by the factor 1.05946, as shown in Ch. 8. We can do this by multiplying the equivalent length of the air column by the same factor, which amounts to an increase in length of 5.946 percent, or approximately 6 percent. Hence, to lower the pitch of the trumpet a semitone, we must make it longer by about 6 percent of 148 centimeters, which gives 8.8 centimeters to be added by the second valve. This makes the new equivalent length 156.8 centimeters. To go down another semitone now requires adding 6 percent of this, or 9.3 centimeters. The first valve must then add 8.8 + 9.3 = 18.1 centimeters to go down two semitones. The equivalent length is now 148 + 18.1 = 166.1 centimeters. The next step down will require 6 percent of this, or 9.9 centimeters; however, the second valve adds only 8.8 centimeters, so the combination of valves 2 and 1 is 1.1 centimeters short of the amount required to lower the pitch a minor third.

The same discrepancies will obviously arise when the third valve is used in combination with the others. Since they are a consequence of the arithmetic involved in calculating percentage increases, they will exist in all the brass instruments that use valves in combination. Various valve systems have been invented in the past to try to solve the valve addition problem; these involved complicated mechanical arrangements whereby pressing two valves at once added an extra bit of tubing over that inserted by the valves individually. Descriptions of these systems may be found in the literature; they are not in common use today.[7]

In practice, therefore, compromises must be made to minimize the discrepancies produced by valve combinations. The valve slides for valves 1 and 2 may be adjusted slightly longer than required, so that the first two semitones are slightly flat. This can be done just enough so that their sum will make the third semitone slightly sharp, but not as badly off as it would be if the first two were in tune. Similarly, the third valve slide may be adjusted to give a shift of somewhat more than three semitones, so that when it is used with the other valves the combinations will be about right. Various tuning schemes have been proposed for getting the best compromise overall.[8]

The discrepancy is worst for all three valves down, as might be expected from the arithmetic. For this reason the modern trumpet has provision for moving the valve slide of the third valve with the left hand little finger while the instrument is being played; by lengthening this slide when all three valves are down, the discrepancy may be removed.

For the smaller brass instruments, the valve problems are not too serious and the player can compensate for them, since his lip has a measure of control over the sounding frequencies of the system. In the trumpet, for example, the playing frequency may be varied over a range of as much as ¾ semitone—from some 40–50 cents below the nominal playing frequency to 30–40 cents above. However, the tone suffers if the correction is too extreme, so it is better to have the instrument as well as possible in tune. The detailed acoustical explanation for the control by the lip of the playing frequency has yet to be worked out.

The lip control is also necessary to compensate for the intonation problems caused by the differences in temperature and composition along the air columns, as described in Ch. 11 for the woodwind instruments.

Instruments: The Trumpet

Over the years a great variety of brass instruments have been designed and built. Some of these were quite outlandish in appearance.[9] Descriptions of the older instruments may be found in the literature.[10] Most of these have now become obsolete, leaving us at the present time with a few well-defined types. All of them are based on the acoustical principle outlined for the trumpet, with a mouthpiece at one end and a flaring bell at the other, and so having resonances that except for the lowest, approximately fit the harmonic series.

We have already described the basic acoustical features of the *trumpet*. Its total length of about 137 centimeters (54 inches) is kept to reasonable proportions by coiling it into an oval with one complete turn. A turning slide is provided at the first U-bend (away from the mouthpiece) to allow tuning of the resonance modes without valves; these are called the *open tones*. The bore diameter of the main tube is about 1.1 centimeters (0.45 inches), tapering down to about 0.9 centimeters (0.35 inches) in the last 12 to 24 centimeters (5 to 9 inches) from the mouthpiece, and opening to a diameter of about 11 centimeters (4.5 inches) at the end of the bell. The instrument has a relatively long cylindrical portion and a relatively short flaring part, the bell starting at about the last U-bend and comprising about one-third the length.

The most prevalent trumpet is the B♭ instrument, but a variety of other sizes have been made and used; those playing in higher keys are useful for playing trumpet parts that lie in the higher ranges. The trumpets are all transposing instruments (with the obvious exception of the trumpet in C).

The *cornet* is very much like the trumpet. A greater proportion of its length is flared, and less of it is cylindrical. It also has an extra turn which reduces its overall length and gives it a somewhat altered appearance from the trumpet; the tone, however, is very little different.

The Trombone

Like the trumpet, the trombone comes in a number of sizes. The commonest instrument of this group is the tenor trombone in B♭; however, it is not written as a transposing instrument. The tube length of the trombone is twice that of the trumpet, so it plays an octave lower. The bore is slightly larger, about 1.3 centimeters (0.5 inches) in diameter. The bell occupies about ⅓ of the length and opens to 15 to 20 centimeters (6 to 8 inches) in diameter.

The bass trombone is a somewhat more complicated version of the instrument. It utilizes an additional length of tubing which may be inserted into the system by means of a single valve and which lowers the pitch of the instrument by a fourth. In addition to extending the range of the instrument downward, the availability of the valve makes some passages easier to play; at the same time it complicates things for the player, because adding the extra section changes the positions of the slide.

The resonance curve for a trombone is shown in Fig. 7. As in Fig. 4 for the trumpet, a relative frequency scale is drawn along the bottom and the second resonance made to coincide with relative frequency 2. The higher ones then very nearly coincide with frequencies 3, 4, 5, and so on, forming an approximate harmonic series. As in the case of the trumpet, the lowest resonance of the trombone does not at all fit this series, but is much too flat. The pedal tone on the trombone is therefore produced by the same mechanism that was outlined above for the trumpet. Thirteen resonances appear in Fig. 7; some twenty more higher resonance peaks were found for the trombone measured, with heights that get progressively smaller as the frequency is increased; these are not shown in Fig. 7. The trombone does not ordinarily play above the eighth resonance.

The French Horn

The *French horn* is based on the same acoustical principles as the other brass instruments, but differs from them in one fundamental respect; whereas the trumpet and trombone play up to about the eighth resonance mode, the horn uses the series up to an octave higher, as far as the sixteenth. This requires that the high resonances of the instrument be pronounced and distinct. The shape of the air column in the horn has been empirically developed to accomplish this; the bore is in the

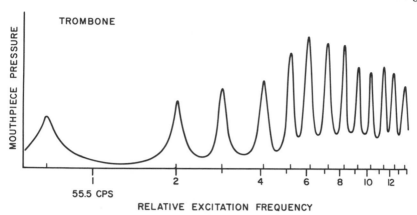

FIG. 7. Resonance curve for a trombone.

form of a long cone with a gradual taper for about two-thirds of its length, after which it flares rapidly to a large bell.

The resonance curve for an F horn is shown in Fig. 8. In contrast to the resonance curve for the trombone, the resonances in the horn are quite pronounced out to the twenty-second, the last one shown. A half-dozen peaks past this are not shown; they diminish rapidly in height. In the horn, the lowest mode more nearly coincides with the fictitious fundamental than it does in the trumpet and trombone.

The length of the air column in the horn is somewhat longer than that of the trombone; to keep its physical length to reasonable propor-

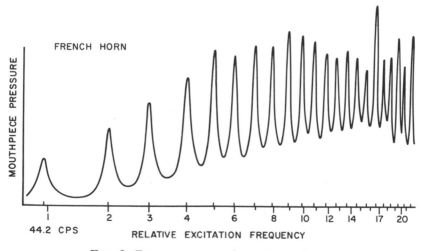

FIG. 8. Resonance curve for a French horn.

tions, it is coiled into a circle. The horn mouthpiece differs from that of the trumpet in having a deeper cup with a more gradual transition to the back bore, as shown in Fig. 3(b). The acoustical reasons for this difference are not established. In the literature, statements may be found asserting the importance of having the stream of air from the player's lips enter the mouthpiece in the proper direction.[11] These statements are illustrated by drawings showing the air stream bounding off the inside of the mouthpiece in various ways and so producing different tone qualities. Such representations are meaningless and should not be taken seriously.

Since the horn plays in a range where the modes are close together, it has more available notes and thus more flexibility than, for example, the bugle. However, it still gives only the notes corresponding to its modes; hence, as originally used in the orchestra, the length of the horn was chosen to give a series of modes whose frequencies were in the key of the composition being played. In practice, this was done by using a single horn together with interchangeable pieces called *crooks*, which could be inserted to change the length of the instrument and put it in the right key. This usually required the horn player to carry several pounds of such crooks to rehearsals and performances.

It was subsequently discovered that if the horn were shaped so that the hand could be placed in the bell, the pitch of the instrument could be lowered by an amount that depended on how far the hand was inserted. Placing the hand in the horn bell restricts its area; this increases the acoustic mass at the end of the resonant air column, as we pointed out in the discussion of the Helmholtz resonator in Ch. 4, and so lowers the frequencies of the system. The technique of hand closure was developed to the point where a complete chromatic scale was playable over part of the instrument's range. However, the hand-closed tones differed in quality from the open tones, and much practice was required to get a reasonably even scale.

The development of valves and their application to horns eased the horn player's problems tremendously. By the use of valves, chromatic tones could be obtained without hand closure, and it was possible to play in different keys without having to interchange crooks. The principle—that of inserting additional tubing—is the same as for the trumpet; however, horn valves are constructed differently, based on rotational rather than sliding action.

The horn in common use at the present time is a double instrument, consisting essentially of the combination of a horn in F and a horn in Bb.[12] The same mouthpiece and bell are used and are switched through two separate channels of different lengths by means of a special valve operated by the thumb. The three main valves each have two sets of windways and two sets of valve slides, one for each channel, so the same three valves can be used for either horn. The length of the F horn is about 375 centimeters (148 inches), $\frac{4}{3}$ times the length of the trombone, so its lowest usable mode is a fourth below that of the trombone, or F_2. In general, the F horn is used for low notes and the B♭ horn for the high notes; here it gives more security to the player, who can attack the notes with less risk of hitting the wrong mode. The horn is normally played with a hand somewhat in the bell for control of tone quality and help in intonation. Since this lowers the pitch, the instrument as normally tuned plays somewhat sharp when open.

If the hand is thrust as far as possible into the bell, a tone of quite different quality is produced, called a *stopped* tone. The pitch apparently rises about a semitone when this is done, so the player must transpose down. There has been considerable argument about the reason for this rise in pitch. It is commonly but erroneously thought that it is because the air column has been shortened by filling part of it with the hand.[13]

What actually happens under hand stopping is an extension of the hand-closure effect described above. The acoustic mass at the end of the horn is increased so much that the frequencies of all the modes are pulled down to where each mode has moved to about a semitone above the original frequency of the one below it. A given note is then played in the next higher mode when the horn is stopped, the one originally used for the note having gone much too flat. This can be demonstrated by blowing a note on the horn while the hand is slowly inserted to the stopping position. If no attempt is made to hold the note to pitch with the lips, the mode can be followed smoothly down in frequency to its final position, below its original value. If the player tries to hold the pitch, the note jumps abruptly to the next higher mode as the hand is inserted, and ends up higher in frequency than its original value. A given stopped note is thus played on a higher mode than the unstopped note.

Other Instruments

Various other brass instruments have been developed and are in use at the present time; descriptions of them may be found in the literature.[14] They are all based on the acoustical principles already outlined. The *tuba* is a familiar example, having a very long air column to produce very low notes. The valve addition problem becomes rather troublesome in the larger instruments: for this reason a fourth valve (and sometimes a fifth) is frequently added to the tuba. It lowers the pitch a fourth, like the valve on a bass trombone, and so substitutes for the combination of valves 1 and 3. It also extends the range of the instrument downward.

Among the brass instruments used in the past which have now become obsolete, there is one worth mention because acoustically it was completely different; this was the *ophicleide*. This instrument was a hybrid; it was blown with the lips on a mouthpiece, like any other brass instrument, but the tube length was altered by opening holes along the length of the tube, as in the woodwinds. At one time there was a whole family of these instruments, of different sizes and pitches. However, they did not survive and are now mostly curiosities.

Power and Tone Quality

The brass instruments are capable of generating considerable sound output. The trombone, for example, can momentarily develop five watts of acoustical power, more than any other instrument outside the percussion group. Played fortissimo, the brass section can almost completely mask the remaining instruments of a symphony orchestra. This is not because the player can put more power into the brass instruments; the actual power expended in blowing them is not much different from that for the woodwinds.[15] Furthermore, the internal standing waves generated in the brass instruments are not too different in amplitude from those produced in the woodwinds. A trumpet blown loudly, for example, will develop a mouthpiece pressure about twice that produced in the mouthpiece of a clarinet blown loudly; this is a difference of only six decibels. Most of the difference in sound output is due to the difference in radiation characteristics of the two families of instruments; as pointed out above, the bell on a brass instrument has a large area, which allows it to radiate sound much more effectively than the small holes on a woodwind instrument.

Another effect of the bell is to give the brass instruments a much more directional characteristic. We pointed out in Ch. 2, in connection with diffraction, that the sound radiating from an opening will spread out in all directions. It will do this practically uniformly if the wavelength of the sound is much larger than the diameter of the hole. This is the case for the woodwinds, in which the tone holes are small compared to the wavelengths of any of the harmonics; the woodwinds will therefore sound much the same in all directions. This will also be true for the low-frequency harmonics of the brass instruments, for which the wavelengths are much larger than the bell diameter. At high frequencies, however, where the wavelengths become smaller than the bell, diffraction effects are much less pronounced and the emerging sound is much more concentrated along the axis of the instrument.[16]

This means that the trumpet will sound considerably different when heard in the direction the instrument is pointing than it will off to one side. The incisive sound of the trumpet in a symphony orchestra is partly due to the fact that the instruments are pointed at the audience, whereas the much more mellow sound of the horns is partly because the instruments are pointing the other way, so that the high harmonics do not as directly reach the audience. Horn players frequently complain (and justifiably) of having to play at the rear of the stage with their instruments pointing toward a curtained backdrop that absorbs most of the sound.

The fact that all of the sound comes out of the bell makes it possible to employ *mutes* with the brass instruments. These are hollow devices inserted in the bell of the instrument; they are constructed according to various designs. A simple form of mute is shown in Fig. 9. The mute

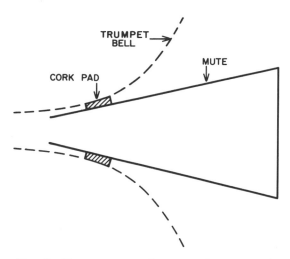

Fig. 9. Cross-section of a simple type of trumpet mute.

generally does not completely fill the bell, being spaced away from it by cork spacers, which also serve to hold the mute in place. Acoustically, the mute is a kind of Helmholtz resonator, as may be seen in Fig. 9. Its resonance frequency is some 200 to 300 cycles per second, although some models of a more complicated structure also have higher resonances. The mute absorbs sound strongly in the neighborhood of its resonances, thus restricting the sound output in this range. This accentuates the high harmonics relative to the low ones, and produces the quality change in the resulting sound. In general, the mute tends to reduce the amount of sound output of the instrument. For some notes, however, the sound output may actually be increased by use of the mute.[17]

The source of the harmonics in the brass instrument tone is not as obvious as it is for the other instruments. In the clarinet, the opening and closing of the reed admits puffs of air; this constitutes a periodic flow of air with a considerable number of high harmonics to excite nearby resonances in the air column. This does not appear to be so for the brass instruments. Observations of the motion of the lips when playing the trumpet have been made; with the use of a transparent mouthpiece, the moving lips may be photographed on moving picture film. Analysis of the pictures obtained shows a rather smooth opening and closing of the lips, as in Fig. 10, the motion being almost sinu-

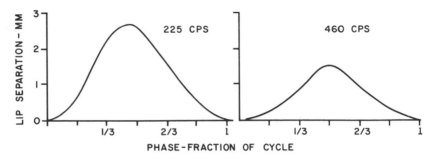

FIG. 10. Motion of the lips during one cycle of a vibration for two different notes on a trumpet.

soidal.[18] There are no abrupt puffs of air into the mouthpiece to generate harmonics of considerable amplitude. It has been suggested that the harmonic structure of the trumpet tone is due to overloading of the air column—that is, the production of such a high sound pressure that the wave is distorted as it travels down the air column and so generates harmonics.[19] This point is not yet settled; it is one of a number of problems relative to the behavior of the brass instruments that needs more investigation.

The material from which brass instruments are made has generally been considered to have an important influence on its tone; in fact, the term "brassy" implies a certain tone quality. However, opinion on this matter has been divided, as is usually the case. One investigator made a trumpet of wood; according to his report, it had a tone that could not be told from that of a brass trumpet when both were sounded behind a screen. At the other extreme, another investigator made a trumpet out of paper; he reported that its tone was "papery."

We should expect the same considerations to apply here as apply to the woodwinds: wall vibrations are present and can be felt, but they should not be expected to radiate enough sound to affect the quality of that normally radiated by the instrument, or to vibrate sufficiently to affect the internal standing wave. Hence we should not expect the material to have any influence on the tone. Experiments on one cornet bear out this conclusion; covering the instrument with putty did not change its tone quality.[20] Since such a coating would stop wall vibrations, it follows (at least for this one instrument) that wall vibrations were not important. However, more work needs to be done to settle this matter for the brass instruments generally.

CHAPTER THIRTEEN

THE PIANO

The *piano* uses vibrating strings as sound sources and is frequently associated musically with the string instruments; acoustically, however, the differences between it and the strings are considerable. All of the instruments discussed up to now have been capable of providing a steady tone, which we have defined as one that can be maintained for a period of time without change in waveform. The piano, however, produces a tone that is not steady, but that instead is continually changing with time, although rather slowly. We might call the piano a "quasi-steady" or "quasi-transient" instrument. Whatever we call it, the musical implications of this nonsteady tone are quite important, as we shall see.

Structure

The essential elements of a piano are shown in Fig. 1. The string (actually a steel wire) is fixed at its two ends on the *frame*. At one end it is wound around the *tuning pin,* which is set into a wooden member

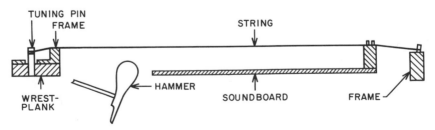

FIG. 1. Elements of a piano. (Not to scale.)

called the *wrest-plank* mounted on the frame; this construction enables the pin to be rotated so as to change the tension on the string and tune it to the right frequency. Near the other end, the string passes over a wooden bar mounted on the *soundboard;* small metal pins driven into the wooden bar hold the wire in position but allow it to slide longitudinally when the tuning pin is adjusted. The string is struck by the *hammer* and set into vibration; the string vibrations are communicated through the supporting wooden bar to the soundboard and radiated into the air.

To produce the right vibration frequency, the tension in the piano string is rather considerable. It will vary depending on the size of the piano, position of the string, and so forth, but can amount to some 150–160 pounds force. Since there are some 230 strings in a piano, the total force on the frame is quite large, amounting to as much as a weight of 30 tons in a modern grand.[1] The frame must obviously be able to withstand this force without altering its shape. In the early days of piano manufacture, the frames were made of wood and (unless musicians were less fussy in those days) the difficulties of keeping a piano in tune must have been considerable. About 150 years ago piano makers started using iron frames for their pianos, with great improvement in the stability of tuning.

Since vibrating strings produce very little sound in the surrounding air, they need some assistance in order to radiate effectively. This is provided for by the soundboard, as shown in Fig. 1, which serves the same function as the body of the violin in making the vibrations of the strings audible. The soundboard is thus a very important part of the piano, and its design involves some of the same kind of problems met with in designing the violin—namely, it should respond well and uniformly to all the vibration frequencies in the range of the piano. The structure of the soundboard in the modern piano has been worked out as a result of many years of trial and error; the acoustical basis for its design is not well established, although some work has been done in this direction.[2]

The strings of a piano are made of steel in order to withstand the tension to which they are subjected. As the instrument developed, it was found that the tone was better if more than one string was provided for each note; present practice is to use three per note over most of the range. For the higher notes in the piano, the strings are made shorter so as not to increase their tension too much. The tension could be kept the same if the string lengths were halved for each octave rise; in practice, this makes the high treble strings too short, so the length ratio per octave is made 1.88:1 to 1.94:1 rather than 2:1. This number is called the *stringing scale* of the piano. (The numbers given are for a grand piano.) The strings are also of smaller diameter for the higher notes.

The stringing scale cannot be maintained in the bass, since the strings would then become much too long; to get the low frequencies in strings of practical length, either the mass must be increased or the tension reduced. Decreasing the tension results in a poor tone quality. Increasing the mass by using a larger single wire also results in a poor quality because of the stiffness of the wire; we will discuss this later. To get the right tension and mass for best tone quality, the bass strings are *overwound*—that is, a steel wire is covered by winding brass or copper wire over it. This increases the mass without adding to the stiffness.

The Piano Action

The piano string is set into vibration by striking it with a hammer, which in turn is set into motion by striking a key. As we might anticipate, this operation is not as simple as it sounds. There are a number of important conditions that must be satisfied:

(1) The hammer must not remain in contact with the string after striking it; if it did, it would stop the vibrations of the string. It must rebound immediately after striking.

(2) After rebounding, the hammer must not be able to bounce off the mechanism and strike the string a second time; it must be held off in some manner.

(3) In order to be able to repeat notes quickly, it should be possible to press the key a second time and activate the hammer without first having to allow the key to return all the way up to its original position. The mechanism should reset itself when the key has moved less than halfway back up.

(4) The tone should persist as long as the key is held down and stop immediately when it is released.

The rather complicated mechanism that does all this is called the *action*. Fig. 2 is a sketch of the action of a grand piano as it exists at present, after some 250 years of development. It works in the following way:

(a) When the key (shown shortened, to keep the figure within bounds) is pushed down on its end (at the left) it rotates about the pivot and lifts the capstan. This causes the wippen to rotate about its pivot and raises the jack, which is pivoted to the other end of the wippen. The upper end of the jack pushes on the roller attached to the hammer stem and lifts the hammer toward the string. Just before the hammer reaches the string, the lower end of the jack strikes the jack

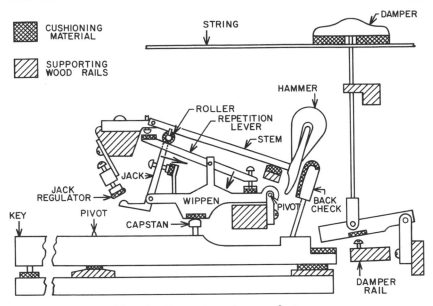

FIG. 2. The action of a grand piano.

regulator; this causes the jack to rotate so that its upper end moves out from under the roller and no longer pushes on it. If the piano key is pressed slowly, the hammer will rise to within about one-quarter inch of the string and then fall back; the mechanism will never press it against the string. If the key is pressed quickly, the momentum the hammer acquires will keep it moving after the upper end of the jack comes out from under the roller, so the hammer strikes the string. At that instant it is completely free from the lifting mechanism, so it will rebound off the string and not remain in contact with it. This satisfies the condition (1) above.

(b) When the hammer rebounds, its lower end is caught by the back check attached to the key; this keeps the hammer from bouncing off the mechanism and striking the string again. This satisfies condition (2).

(c) When the key is now lifted somewhat, the roller rests momentarily on the repetition lever, which is pulled by a spring (not shown) in the direction of the arrow. This allows the upper end of the jack (also pulled by a spring in the direction of its arrow) to slip back under the roller. Now if the key is pressed again, the hammer will be thrown against the string as before; the mechanism thus resets itself when the key has moved back up less than halfway. This provides for the rapid repetition of notes and satisfies condition (3) above.

(d) When the key is pressed, the damper is raised off the string; this allows the vibrations to last until the key is released, when the damper descends and stops them. This takes care of condition (4). If desired, the dampers on all the strings may be raised independently of the keys, by means of the damper rail running under all the dampers and attached to a pedal operated by the right foot.

This complicated mechanism, needed for the proper production of one note, must now be repeated for each of the 88 notes on the piano keyboard. The total number of parts in a piano is therefore quite considerable.

In the grand piano, the entire action is arranged so that it can be moved a short distance to one side by means of a pedal (the *soft pedal*) operated by the left foot. This shifts the hammers so that they strike two strings instead of three, and the tone is softer and of different quality. In early pianos the hammers struck only one string when this pedal was used, hence the term "una corda" still seen in piano music.

In the upright piano, the action is of a different design, but must still conform to the same conditions outlined above. The difference in design produces a somewhat different "feel" in the upright. The soft pedal in the upright does not move the action as it does in the grand; instead, it shifts all the hammers closer to the strings.

Factors Affecting Piano Tone: Hammers

The hammers that strike the piano strings are formed of heavy felt glued to a wooden body. The tone they produce depends on the shape of the head of the hammer. It also depends to a considerable extent on the condition of the surface of the felt. If the surface is hard, the higher harmonics of the string vibration are emphasized and the tone is brighter; the piano tuner can produce this condition by treating the surface of the felt with a hot iron. Conversely, if the surface is made soft by pricking it with needles, the higher harmonic content is reduced and the tone is softer. In this way the tuner can vary the tone of the piano to a considerable extent.[3]

The tone also depends on the point along the string where the hammer strikes. We saw in Ch. 3 that the amplitudes of the harmonics in the vibration of a plucked or struck string depend on the point of striking; if this point is a node for a particular harmonic, that harmonic will not sound. The hammers in the piano are arranged to strike the string about one-seventh to one-eighth the way along it. It is frequently stated that this is done to reduce the amount of seventh harmonic in the tone, this harmonic being assumed to be detrimental. Helmholtz seems to have been the first to make this statement, and it has been echoed since by many others.[4] Actually the hammers do not strike the string at a point, but are in contact with a certain length of string at the instant of striking. As a result, the above condition for the non-production of harmonics does not apply, and measurements show that the seventh harmonic is present in the piano tone with as much amplitude as the neighboring ones.[5] The point along the string where the hammer strikes has been chosen as about one-seventh the string length because this is where it produces the best sound.

Tuning of Unisons

Once the hammer has struck the string and rebounded, the string continues vibrating with an amplitude that depends on the speed of the hammer at the instant it strikes. If the key is held down so as to hold the damper off the string, the vibration gradually dies away as the energy imparted to the string by the hammer is used up. Measurements show that about half the energy is dissipated by air friction and half by transmission to the piano sounding board.[6] It is also found that the decay rate of the sound from the piano string is actually double; that is, the string vibration decreases rapidly immediately after the string is struck and then more slowly thereafter. This double decay rate is characteristic of conventional piano tones. If the strings are tuned to give a single decay rate, the tone is not as good.

Since the piano is provided with three strings per note over most of its range, its tone will obviously depend on how the strings are tuned in relation to one another. Rather surprisingly, it has been found that if the strings are tuned to precisely the same frequency so as to be in exact unison, the tone is not good.[7] Such exact tuning gives the single decay rate mentioned above, and results in the tone disappearing too fast. If the strings are put slightly out of tune, the tone lasts longer and sounds better. Listening tests made on recorded piano tones showed that the preferred tuning was one for which the strings were spaced one to two cents apart in pitch.

There is a considerable amount of noise in the piano tone. Some of it comes from the finger striking the key and some from the hammer hitting the string. The various parts of the action, suddenly set into motion and suddenly stopped, contribute their share, even though cushioned as much as possible by pieces of felt. The noise part of the tone is generally disregarded by the listener. However, if piano tones are synthesized (as described below) without the usual amount of noise, the difference is quite apparent.[8]

Inharmonicity

Up to this point, we have assumed the piano string to be the simple string discussed in Ch. 4, which we assumed to be perfectly flexible. If the simple string is pulled out of its normal position, the restoring force tending to pull it back again arises only from its tension. However, if the string has stiffness, an additional restoring force appears because the string is bent when it is displaced from its normal straight-line position. This is the case for the actual piano string, which, being made of steel, has a considerable amount of stiffness. This modifies its behavior in an important way. When the string is displaced, the additional restoring force due to stiffness results in an increased acceleration of the portions of the string and produces a higher vibration frequency for all the partials of the string vibration. The additional restoring force provided by the stiffness increases more rapidly with the curvature of the string than does that provided by the tension; this results in the frequencies of the higher partials of the string being increased more than the frequencies of the lower partials. Calculations show that the increase in frequency depends on the square of the frequency; that is, the octave will be shifted four times as much as the fundamental.[9]

The practical result of all this is that the partials in the tone of the actual piano string are not integral multiples of the fundamental frequency, and so are not strictly harmonic. Instead, the higher partials become progressively sharp with respect to the harmonic frequencies. Measurements on one piano, for example, showed that the frequency of the fifteenth partial was very nearly sixteen times the fundamental frequency.[10] This departure from the harmonic relationship is called the *inharmonicity* of the partials. It is possible in the piano because the string vibration is not a steady vibration, but is continually changing with time.

The inharmonicity of the partials of the tone of the piano string has a pronounced effect on its quality; in fact, it is an essential part of what we think of as piano tone. This has been demonstrated by listening tests made on synthetic tones.[11] Such tones may be built up by combining the outputs of electronic oscillators (to be described further in Ch. 15) whose frequencies can be set at any desired values and then arranging (again electronically) for this output to have the same rise and decay characteristic of an actual piano tone. Since the various characteristics of such synthetic tones may be easily altered, the effect of such alterations on the quality of the tone can be judged. If tones are built up of partials that have the same strength and the same decay rates as those of the piano string but are strictly harmonic, the tone does not sound like that of a piano. Conversely, if the partials are made inharmonic to the proper degree, the synthetic tones cannot be told from real piano tones.

Although a certain amount of inharmonicity appears essential, too much is undesirable. The inharmonicity in a given string depends on its length, the amount increasing as the string is made shorter. It has been found that the bass tones on a piano are better if the strings are made long; this accounts for the length of the modern concert grand. Such strings will have a smaller amount of inharmonicity than the short strings on a small piano, so this appears to be a desirable condition. The reason for this may be related to the circumstance that the fundamental is practically absent in low piano tones and must therefore be supplied by the ear as either a difference tone or a periodic waveform change, as described in Ch. 6. The presence of inharmonicity means that the difference frequency between adjacent partials in the tone is not a constant amount equal to the fundamental frequency, but instead will be larger for higher partials. Too much inharmonicity will then result in a difference tone that is poorly defined and not acceptable to the ear.

It has been calculated theoretically that a small mass placed at the proper position near one end of the piano string could reduce the frequencies of the partials enough to make them very nearly harmonic.[12] Apparently no one has as yet tried this with an actual piano. From our discussion above, it appears that such an alteration would produce a considerable change in a tone of a piano. It might not be a desirable change, at least in terms of our present conditioning to pianos as they exist, but it would be an interesting experiment.

Tuning the Piano

Instruments to be played in ensemble will obviously sound best if they are all tuned to the same scale, and in Ch. 8 we described how the tempered scale was developed as a workable compromise between the various conflicting factors. The brass and woodwind instruments can be adjusted somewhat in pitch while played, but they should be built to the tempered scale so as not to impose too many intonation problems on the player. The keyboard instruments such as the piano and organ must be tuned in this scale, or at least reasonably close to it, since they cannot be adjusted while played.

To tune a keyboard instrument to the tempered scale is a matter of applying arithmetic to the observation of beats, which we defined in Ch. 3. The inharmonicity of the partials of the piano tone causes complications, so let us for the moment discuss the tuning of a keyboard instrument for which each key produces a single steady tone having a reasonable number of harmonic partials. This would be the case for the pipe organ, for example.

The first step is to tune A_4 on our instrument to 440 cycles per second, since that is the standard for our tempered scale. We can do this by sounding this note together with a note from a standard tuning fork. If the two frequencies are not the same, beats will be heard; the tuning adjustment on A_4 is changed until the beat disappears. This note will then be in tune.

Next, we may start with A_4 and set D_4 a tempered fifth below it. If these two notes were set a just fifth apart, the frequency ratio $D_4:A_4$ would be exactly 2:3, so D_4 would be $\frac{2}{3} \times 440 = 293.33$ cycles per second. The frequencies of the harmonics would then be as follows:

A_4: Fundamental, 440.00 cps; second harmonic, 880.00 cps.
D_4: Fundamental, 293.33 cps; third harmonic, 880.00 cps.

We see that the second harmonic of A_4 coincides with the third harmonic of D_4, so these two notes can be tuned a just fifth apart by adjusting D_4 until there is no beat between these harmonics.

What we want, however, is a tempered fifth, which is narrower than the just fifth. Referring to the table of tempered scale frequencies given in Ch. 8, we see that D_4 is actually 293.66 cycles per second. This puts the harmonics out of tune, as follows:

A_4: $440.00 \times 2 = 880.00$ cps for second harmonic,
D_4: $293.66 \times 3 = 880.98$ cps for third harmonic,

Difference: 0.98 beats per second.

Hence we will get 0.98 or very nearly 1.0 beats per second between these harmonics when D_4 and A_4 are a tempered fifth apart. The interval can thus be set by first adjusting D_4 to get no beat, then narrowing the interval by raising D_4 just enough to get 1.0 beats per second. The reason for getting the no-beat just interval first is that we then know which way to go; it is also possible to get 1.0 beats per second when the interval is wider than the just value, that is, when D_4 is 293.00 cycles per second.

With D_4 as a new starting point, we can now set G_4 a tempered fourth above by listening for beats between the fourth harmonic of D_4 and the third harmonic of G_4, as follows:

G_4: $391.99 \times 3 = 1175.97$ cps for third harmonic,
D_4: $293.66 \times 4 = 1174.64$ cps for fourth harmonic,

Difference: 1.33 beats per second.

Since the tempered fourth is slightly wider than the just fourth, we would set G_4 to first get no beat and then widen the interval by raising G_4 slightly to get 1.3 beats per second.

The note C_4 may now be set a tempered fifth below G_4 in the same way; the arithmetic now gives 0.88 beats per second between the third harmonic of C_4 and the second harmonic of G_4. Similarly, F_4 may now be set from C_4, and so on. Each fifth and fourth has its own beat rate, depending on its position; going up an octave would obviously double the beat rate, since it doubles all frequencies. Once these have been worked out, all the notes in a given octave may be set by properly adjusting the beats. Tables are available for this purpose, giving beat rates for fifths, fourths, thirds, and sixths over a range of notes of the keyboard.[13]

Once a given octave has been tuned, the octave above may be tuned to it. This is done by listening for beats between a given note and the second harmonic of the note an octave below. When the higher note is adjusted so the beat disappears, the octave is in tune. In the same way, lower octaves may be tuned.

To apply this process to the piano, the tuner first inserts a strip of felt between the piano strings so that only one string of each note can sound. He then tunes one octave—say, F_3 to F_4—to equal temperament, a process called *laying the bearings*. In the course of working back and forth by fifths and fourths, small errors can inevitably accumulate; to minimize these, certain cross-checks are available. For example, starting with C_4 (set to a tuning fork), he can set F_3, G_3, D_4, and A_3 in four steps, using fifths and fourths. The major third F_3A_3 can now be checked; if it is correct, there should be 7.0 beats per second between the fifth partial of F_3 and the fourth partial of A_3. Similarly, the rest of the octave can be tuned and checked. Higher and lower octaves can then be tuned; various checks to minimize errors are available here also.[14]

With one string for each note properly tuned, the remaining two strings can be tuned to it, one at a time, by getting approximately zero beat for the pair. This setting cannot be too exact, because of the desirability of having a slight mistuning between strings, as described above. However, a difference of two cents between two strings for C_4 would give about one beat in three seconds—about as slow as can be judged as a beat. Hence in practice the tuner sets for as close to zero beat as he can, and the unavoidable small remaining tuning errors provide the desirable mistuning.

Once tuned, a piano does not stay that way. Since the forces acting on the frame, wrest-plank, and sounding board are quite large, these members will "give" over a period of time. As a result, the piano will go flat and must be periodically retuned.

Since so much of the piano is wood, the relative humidity also has an effect on the tuning, raising and lowering it depending on the amount of water vapor in the air.[15]

Effect of Inharmonicity on Tuning

The tuning process described above is based on the assumption that the partials in the piano are harmonic. Since this is not quite true, we should expect the inharmonicity to affect the tuning. For example, if an octave is tuned by getting zero beat between the second partial of the lower tone and the fundamental of the higher tone, the interval so obtained will be too wide if the second partial is sharp as compared to the second harmonic. Since the partials in the piano tone get progressively sharper as we ascend the series, we should expect the higher octaves in the piano to be sharp and the lower ones flat with respect to the center octave.

This is the case for pianos as actually tuned by expert tuners. Measurements made with the stroboscopic frequency meter show that well-tuned pianos generally follow a tuning curve like that shown by the solid line in Fig. 3, which is an average for sixteen pianos.[16] Individual pianos will of course show deviations from this curve. Pianos with long strings will show less deviation from the tempered scale values, particularly in the bass. For example, the points to the left of the solid line in Fig. 3 are for the bass notes of an upright piano whose longest string was 1.03 meters. The points to the right of the solid line are for a concert grand whose longest string was 2.02 meters.

There will also be deviations from the curve of Fig. 3 that depend on the individual tuner, but on the average the highest notes may be more than 30 cents sharp and the low ones 30 cents flat as compared to their expected frequencies. A tuning of this kind is said to be *stretched*. Since it is used in the piano, it is not strictly accurate to say that the piano is tuned to the tempered scale.

It appears that the stretched tuning can be sufficiently accounted for by the inharmonicity of the partials.[17] Whatever the cause, we are quite conditioned to such tuning; listening tests show that stretched tuning is unequivocally preferred over unstretched.[18] A piano accurately tuned to the tempered scale over its entire range sounds quite flat in the upper register.

Touch

Pianists have used the term *touch* for a long time and over the years it has acquired an almost mystical connotation; a pianist is praised or condemned depending on the adjective modifier used. He may have a "singing," "beautiful," "pearly," or "velvet" touch—the list is endless, depending only on the imagination of the favorably inclined critic. Conversely, if the critic is in a bad humor, the pianist's touch may be "harsh," "percussive," or whatever. The worth of a pianist is measured by the quality of his touch.

Implied in all this is the belief that the pianist can in some manner control the quality of the tone he produces from the piano string by the way he strikes the key. Books have been written asserting this as fact.[19] It has been stated categorically that if the piano key is put in motion suddenly by striking the key with the finger, the tone will be harsh and

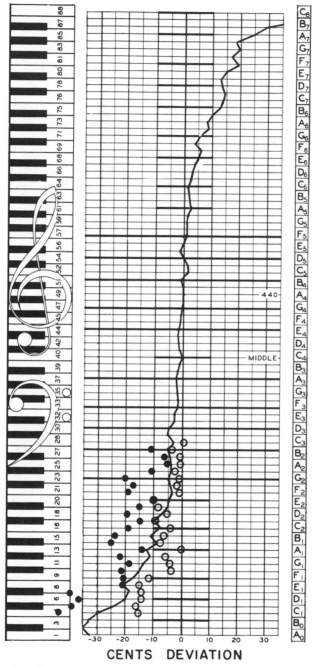

CENTS DEVIATION

FIG. 3. Stretched tuning on the piano; average of sixteen pianos.

strident; conversely, if the key is gradually put in motion by being gently pressed, the tone will be smooth and mellow. If this is true, it follows that much practice would be necessary to acquire the proper manner of depressing the piano keys.

To put it bluntly, this is nonsense. In our earlier discussion of the action of the piano, we saw that at the instant the hammer strikes the string, it is completely separate from the impelling mechanism attached to the key. The speed of the hammer on striking the string depends on how the key is pressed, and determines the loudness of the resulting tone. It also determines to a certain extent the quality of the tone; a loud tone will have a greater number of higher partials than a soft tone, and so will be "brighter," or perhaps "harsher." A given hammer speed will thus produce a certain loudness of tone and with it a certain quality of tone, and the two are not independent; if the loudness is the same, the quality is the same. It does not matter how the hammer attained its speed, whether via a sudden acceleration by striking the key or a slower acceleration by pressing the key; a given final speed will always produce the same tone. It follows that the pianist cannot independently control the quality of the tone of a single note on the piano by the manner in which he strikes the key; a given loudness will always result in a tone of quality corresponding to that loudness.[20]

A detailed investigation of this matter has been made in the laboratory.[21] A mechanical striker was constructed that could depress a key of a piano and impart to it accelerations that could be varied to correspond to different ways of depressing the key with the finger. The speed of the hammer at the instant of striking the string could also be measured. The tone produced could be evaluated by recording its waveform on photographic film by means of an oscillograph. It was found that the waveform varied with the speed of the hammer; however, if the speed were kept the same, then in all cases the waveform was the same, regardless of the kind of acceleration used in striking the key. Furthermore, the waveform produced by a concert pianist striking the key could be duplicated precisely by adjusting the mechanical striker to produce the same hammer speed.

We must conclude that as far as single tones on the piano are concerned, the player does not have the ability to control quality in the manner that has been commonly assumed. The pianist himself may be subjectively convinced that he is doing so, and the adjectives applied by equally subjective critics may convince others that he is doing so. However, the objective listener will be unable to detect these supposed differences in quality by listening to individual piano tones.

Pianists as a group seem remarkably resistant to this fact, which has been pointed out to them for almost half a century. This is strange, since one would think that it would make the pianist's work easier. It does not mean that all pianists sound alike; quite the contrary. The whole concept of piano touch involves a complicated mixture of things: evenness of playing, degree of legato, ability to emphasize an individual note in a chord, and many others.[22] Learning all these is difficult enough without adding objectives that are impossible to realize. (The author is a pianist.)

FIG. 4. Ambiguous writing for the piano; (a) and (b) will sound the same if played with the same loudness.

As a corollary, the kind of piano writing often seen, in which notes are written as very short or staccato while the damper pedal is marked as being held down, is redundant and ambiguous. The two examples of Fig. 4, for instance, literally interpreted, will have identical sounds, and composers may as well save themselves the trouble of writing rests in the music where the pedal is used. This will also spare the pianist the problem of deciding what the composer means by such ambiguous writing.

CHAPTER FOURTEEN

THE PERCUSSION INSTRUMENTS

We finally come to those musical instruments with tones that are completely transient, dying out more or less quickly once they are produced. These instruments are generally sounded by striking them so they are termed *percussion* instruments. The piano is actually a percussion instrument, but one whose tone changes sufficiently slowly to allow it to be almost classified as a steady-tone instrument. The majority of percussion instruments have tones that change more or less rapidly with time. Such tones will have waveforms that differ from one cycle to the next, and so will not be periodic. Their partials will therefore not generally be harmonic and their pitches may be more or less indefinite. Such tones are produced by complicated vibrating systems such as membranes, bars, and plates. We have already discussed some of these in Ch. 4; here we will consider those important to the percussion instruments.

Vibrating Membrane Instruments: The Timpani

An important group of percussion instruments utilizes the vibrations of a thin stretched membrane that is attached to a circular frame. The *timpani* are prominent members of this group. The instrument consists of a membrane, called the *head,* attached to a circular metal hoop and mounted on a hemispherical copper bowl. The appearance of this bowl is responsible for the name *kettledrum* that is also applied to the instrument. The head is set into vibration by striking it a few inches from one edge with the *stick,* usually a wood shaft with a ball of felt at-

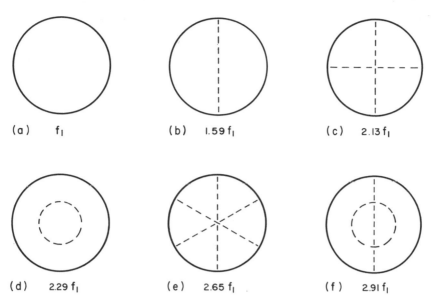

(a) f_1

(b) $1.59 f_1$

(c) $2.13 f_1$

(d) $2.29 f_1$

(e) $2.65 f_1$

(f) $2.91 f_1$

FIG. 1. Modes of vibration of a circular membrane.

tached to one end. Tones are produced either by single blows of the stick or by rapidly repeated alternating blows with two sticks, producing a *roll*.

A few of the modes of vibration of a rectangular membrane were illustrated in Ch. 4; some of these modes were harmonic, but most were not. Some of the modes of vibration of a stretched circular membrane are shown in Fig. 1, together with their frequencies relative to the frequency f_1 of the lowest mode. It will be seen that they are all inharmonic. The head is assumed to be stretched equally in all directions.

The frequency f_1 of the lowest mode is given by a formula very like that of Eq. (3) in Ch. 4 for the lowest mode of vibrating string; that is,

$$f_1 = \frac{0.766}{D} \sqrt{\frac{T}{\sigma}} \qquad (1)$$

where T is the tension stretching the head in newtons per meter width, σ is the density of the head in kilograms per square meter, assumed to be the same everywhere, and D is the diameter of the head in meters. This is the fundamental mode for the head, and is shown in Fig. 1(a). In this mode, all parts of the head are vibrating in phase except of course for the outside edge where the head is supported. The next mode is shown in Fig. 1(b); it has one nodal line across the head, which divides it into two vibrating segments. The frequency is $1.59 f_1$.

The next mode has two nodal lines, as shown, and a frequency $2.13f_1$. Above this is a mode with a nodal line in the shape of a circle, and a frequency $2.29f_1$. Two more modes together with their frequencies are shown in Fig. 1; higher frequency modes exist but are not important.

When the head is struck with the stick, it vibrates in all its modes simultaneously, just as the string does when it is plucked. However, the fundamental mode displaces a considerable amount of air when it vibrates, since the whole head moves in phase. It therefore radiates sound energy rapidly and is heavily damped, so it dies out very quickly. We may illustrate this by striking the head in the center; a very short "dead" tone is obtained. Since striking the center produces the greatest amplitude of fundamental, it follows that the fundamental makes very little contribution to the timpani tone. It is part of the initial "thump" heard when the head is struck, but not a part of the subsequent sound that determines the pitch and quality of the tone.

The higher modes are not heavily damped in this way. For these modes the head vibrated in segments, with two adjacent segments always a half-cycle out of phase—that is, moving in opposite directions. The adjacent segments therefore tend to cancel one another's sound radiation, so the higher modes do not lose energy rapidly. The lowest frequency that lasts for any time is then that of the second mode, shown in Fig. 1(b). If we take this to be the "fundamental," the higher modes have frequencies that are 1.42, 1.53, and so on times this frequency. Each of these modes radiates sound of its frequency, so the partials in the tone of the timpani are obviously not harmonic. This produces some indefiniteness of pitch in the sound of the instrument.

The assumptions of uniform tension and density of the timpani head that were stated above had to be made in order to work out mathematically the frequencies of the vibration modes. However, these conditions will not necessarily be true in practice; the head may be stretched differently in different directions and, since it is frequently the skin of an animal, it may not have a uniform density. The actual vibration frequencies will then depart more or less from those given above. It is quite possible that by making the tension and density nonuniform in some particular manner, the mode frequencies could be shifted to become more nearly harmonic; this might result in better tone. This matter should be investigated.

A pair of timpani of diameters 71 centimeters (28 inches) and 64 centimeters (25 inches) will cover the range F_2 to F_3. Smaller and larger models are available to go to higher and lower pitches. The instrument is tuned by changing the tension in the head; this changes the vibration frequency, as shown in Eq. (1). The tension is changed by means of six handles around the periphery that exert more or less force on the ring to which the head is attached, and so stretch the head more or less tightly across the kettle. With this arrangement it is not possible to change the tuning quickly, so the modern instrument has a foot pedal connected to all six handles in such a way as essentially to work them all simultaneously. This arrangement permits rapid tuning.

Timpani heads are commonly made of calfskin. This material is quite susceptible to changes in humidity, loosening up in damp air and tightening up when the air is dry, so the player never knows how long his instruments will stay in tune. Plastic timpani heads, which are not affected by humidity changes, are now available; however, for reasons not yet known, their tone is not the same as that produced by calfskin heads and not as well liked by most players.[1]

The kettle on the instrument keeps sound from radiating from the under side of the head and so makes the tone louder. A head uncovered on both sides would radiate sound of equal amplitudes but opposite phases from the two sides. These would effectively cancel each other, and the head would radiate less sound than it does when the kettle covers one side.

The effect of the shape of the kettle on the timpani tone has not been ascertained. Arguments by timpanists on this question, fortified with diagrams showing assumed paths of sound waves reflecting from the inside of the kettle, have no acoustical validity.[2]

Drums

The remaining members of the drum family do not produce sounds of definite pitch. They generally have two heads mounted on a cylindrical body; this construction allows the vibrations on one head to be transmitted through the air inside the drum to the other head and cause it to vibrate also.[3] As a result, the vibrations of the system are more complex than for the timpani, and the tone produced is completely indefinite in pitch. The *snare drum* has wires stretched across one of the heads; these beat against the head when it vibrates and produce high-frequency vibrations that add more noise to the sound of the drum.

Vibrating Bar Instruments

Another group of percussion instruments is based on the vibrations of bars. We discussed these in Ch. 4 for bars fixed rigidly at one end, as applicable to the tuning fork. For bars free at both ends, the vibration patterns of the first three modes are as shown in Fig. 2. They resemble somewhat the modes of vibration of an air column open at both ends, since there are antinodes at the free ends. However, the frequencies are not as simply related. The frequency of the fundamental mode, shown in Fig. 2(a), depends on the length of the bar, being inversely propor-

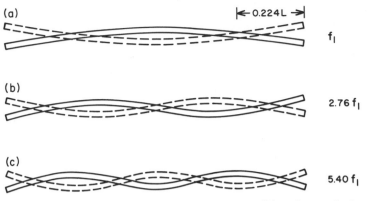

FIG. 2. First three modes of vibration of a stiff bar free at both ends.

tional to the square of the length; we do not need the exact formula. The frequency is independent of the width of the bar. The frequencies of the next two modes, shown in Fig. 2(b) and (c), are respectively 2.7 and 5.4 times the fundamental and thus are not harmonic.

A set of *orchestra bells* consists of a number of bars of graded lengths mounted on a supporting framework; the bars are in contact with the framework only at the nodal points shown in Fig. 2(a) so the vibration of the fundamental is not appreciably affected. The bars are struck near the center with hard mallets. This point is a node for the second partial, so it will be produced with small amplitude. Furthermore, the second and higher partials do not have nodes at the points where the bar is supported, so they are rapidly damped. For this reason, the tone produced is relatively pure. The vibration of the bar lasts for some time after it is struck, but continually diminishes in amplitude as the energy given it is used up by sound radiation and by internal friction in the material of the bar.

The *xylophone* is similar in construction to the bells, but uses bars of wood instead of metal. Wood is a material with a greater amount of internal friction than metal, so the vibrations of a bar of wood are more quickly damped by conversion of the vibration energy into heat. This is the reason for the short, "dry" sound of the xylophone. The bars have placed under them metal tubes closed at one end, of graduated lengths. Each bar is thus supplied with an air column resonant at its fundamental frequency; this helps in the radiation of sound.

Vibrating Plate Instruments

Vibrating plates are more complicated than bars because they have two important dimensions instead of one. As a result, a plate can vibrate in a great many possible nodes with frequencies spaced closely together, as we saw for the top and back plates of the violin. Because of these many closely spaced frequencies, the pitch of a tone produced by striking a plate is generally completely indefinite. The orchestra *cymbals* are an adaptation of vibrating plates. They are usually used in pairs, being struck together, although they may be used singly and struck with sticks of various kinds to get sounds of various qualities. The sound of the cymbal, being a mixture of many frequencies, is essentially what we defined in Ch. 6 as noise, but it will have a larger or smaller proportion of high-frequency components depending on the hardness of the stick used.

CHAPTER FIFTEEN

THE ELECTRONIC PRODUCTION OF SOUND

The tremendous technological developments of the present century have greatly altered our social and economic environment. The field of music has not escaped the impact of this technology; developments in that area of the field of communications known as *electronics* have had a profound influence on the recording and reproducing of music and the production of musical sounds, and show portents of affecting the future creation and performance of music as well. A discussion of the status of present-day electronic and technological developments as they affect music may give some indication of the direction of future developments.

Mechanical Recording and Reproduction of Sound

The first impact of technology on the field of music came with the development of means of recording sound. The preservation of sound on records was made possible by Edison's invention of the *phonograph* about the year 1877. It utilized the conversion of sound motion to mechanical motion in a way that could be reversed, so that the sound could be re-created again.

The earliest version of the phonograph is shown schematically in Fig. 1(a). A cone open at its large end was closed at the small end by

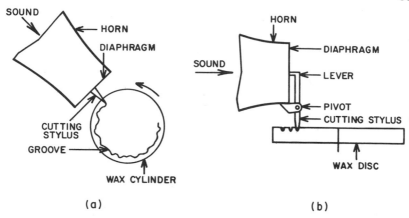

FIG. 1. Mechanical recording of sound. (a) Edison's original method using cylindrical records. (b) An improved method using disc records.

a thin diaphragm. The diaphragm had attached to its center a stylus, arranged to cut a groove on the surface of a cylinder of wax rotating under it. As the cylinder rotated, it was moved along its axis, so that the groove cut by the stylus took the form of a long spiral. Sound waves traveling down the cone produced a varying pressure on the diaphragm, causing it to move in and out; this varied the depth of the groove cut in the cylinder. The process could now be reversed; the wax cylinder could be rotated with the stylus riding in the groove so that variations in the groove depth would cause it to move back and forth and so move the diaphragm. The motion of the diaphragm would then produce sound, which would radiate from the horn as a somewhat limited facsimile of the original sound used in cutting the groove. Since the records made by the process were cylindrical and awkward to store, an improved version was soon developed that allowed the use of flat discs for records.

This version, which in principle is still in use today, is shown in Fig. 1(b). The groove cut by the stylus is on the surface of a flat disc of wax rather than on a cylinder; as the disc is rotated, the stylus is moved across it to make a long flat spiral groove. In the original version, the motion of the diaphragm produced by sound waves coming down the horn was transmitted to the cutting stylus by the lever mechanism shown in the figure. This resulted in a motion of the stylus across the face of the disc instead of up and down into it; the sound was consequently recorded as a lateral displacement of the groove rather than as a variation in its depth. This process could again be reversed, as in the case of the earlier model, to reproduce (with some limitations) the original sound making the record.

In actual practice the wax record made in this manner was not used to reproduce the sound directly, since such a soft material, easily cut and hence suitable for making a record, would not last long when played to reproduce the sound. Instead, a method was developed for copying the records that used a process called *electroplating*, by which metal may be electrically deposited on the surface of the wax record to completely cover it and fill in the grooves. This process produced a copy of the original but in reverse, grooves becoming ridges, and conversely. This metal copy could now be used as a mold and from it could be pressed a large number of copies of the original wax record. These would be made of a material more suitable for the wear encountered in playing.

The mechanical recording and reproducing of sound in this fashion was subject to serious difficulties. As we have seen, the energy in a sound wave is quite small and the amplitude of the motions of the air molecules is therefore extremely minute. For example, in a fairly loud sound, of level say 80 decibels, the amplitude of the motion of the air molecules is only about 10^{-3} centimeters. This will be the order of magnitude of the displacement of the record groove, and it is quite small; this makes it difficult to get a reproduced sound loud enough to be well above the inevitable noise produced by small surface irregularities in the record groove. Because of this, when recording a musical composition, it was necessary to seat the musicians as close as possible to the recording horn, sometimes under conditions of considerable crowding and discomfort, to obtain sufficient sound amplitude. Furthermore, resonances in the horn and in the diaphragm produced various distortions in the recorded sound as compared with the original; high frequencies were particularly difficult to record. The resultant distortions are responsible for the "tinny" sound of the early recordings, noticeable on some of the recently reissued copies.

The development of the technology of electronics and its application to the recording process made it possible to overcome many of these difficulties and to produce the excellent recordings available today.

Electricity and Electronics

To appreciate the various ways in which electronics has affected the recording, performance, and creation of music, it will be helpful to understand some of the simpler concepts of electricity. We do not need to know the behavior of electronic circuits in detail (even engineers have trouble sometimes); the barest essentials will be sufficient.

All matter is composed, in part, of something called electricity. This "stuff" (substance is hardly a correct term) has been the subject of study for many centuries. As a result of this study, we are able to describe quite well how electricity behaves; in fact, its behavior can be described by some of the most beautiful and elegant mathematics in the whole field of physics. Electricity can flow in metallic wires called *conductors,* producing an *electric current,* just as a current of water can flow through a pipe. The electric current is caused to flow by a sort of an electrical pressure called *voltage,* just as in the pipe the water is forced to flow by a difference in the water pressures at the ends of the pipe. The relationship between the flow of current and the electrical pressure is actually simpler than the corresponding relation between the flow of water and the water pressure.

Electric currents flowing in conductors can do various useful things. They may generate heat, for example, the amount depending on the nature of the conductor; in some cases the heat is enough to make the conductor incandescent so that it produces light. Currents can exert forces on one another, which means that they can do work; such work is done by an electric motor. Both the production of heat and the production of mechanical work require that somewhere else some source is available from which to obtain the necessary energy.

Small amounts of such energy may be obtained from *batteries,* which store it in chemical form and can convert it at a steady rate into electrical power as needed. Batteries are useful for operating flashlights and various portable electrical devices, but their outputs are limited. For larger amounts of electrical power, we rely on the local power company, which by means of generators converts the mechanical power produced by falling water or steam turbines into the electrical form that it sends to us along conducting wires.

The essential details of the system that produces electrical power and uses it up again are shown in Fig. 2. This system is called a *circuit,* since the electric current circulates in it as long as a continuous closed path is available. The source of electrical power—the battery or the generator in Fig. 2—produces the voltage in the circuit which causes the current to flow. The current then expends this electrical power by producing mechanical power in the motor or by producing heat (and some light) in the electric light.

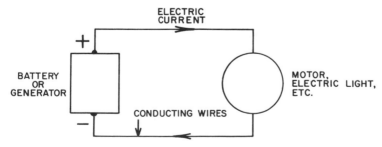

FIG. 2. The basic electric circuit.

The forces between electric currents are of a rather special character. A current can exert a force on another one even though they are separated by some intervening distance, and even if there is no material substance between them to transmit this force. This state of affairs has given rise to the concept of a *magnetic field*. We have all seen and perhaps played with magnets, which are bars of material with ends, called *poles,* that attract and repel one another in curious ways. A small magnet, suspended so it can rotate, will point in a general North-South direction, indicating that the earth itself is a magnet; the *compass* so produced was one of the first practical applications of electricity and has been used for centuries.

The fact that magnet poles exert forces on one another even when separated is pictured in terms of the magnetic field. The space around a magnet is somehow different from ordinary space, being filled with this magnetic field; the field in turn exerts forces on other magnets brought into it. The field around the magnet may be represented by drawing lines, as in Fig. 3(a), although we must remember that the lines have no actual physical existence. Magnet A is producing the magnetic field; this field is exerting the forces shown by arrows on the poles of magnet B. The converse is also true: magnet B has a magnetic

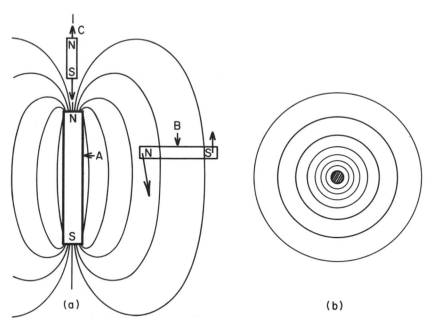

FIG. 3. (a) Magnetic field around a magnet, and the forces on magnets nearby. (b) Magnetic field around a wire carrying an electric current, viewed end on.

field (not shown) that will exert forces on the poles of magnet A. A piece of iron brought into the magnetic field near a magnet becomes a magnet itself, and so is attracted. Such a piece of iron is shown by C in Fig. 3(a). Depending on the nature of the iron, it may lose this magnetism when removed from the field, or it may keep it.

An electric current is surrounded by a magnetic field, as pictured in Fig. 3(b). This is demonstrated by the fact that a current will exert a force on a magnet brought nearby. The converse is also true: a wire near a magnet experiences a force, as shown in Fig. 4(a). At the pres-

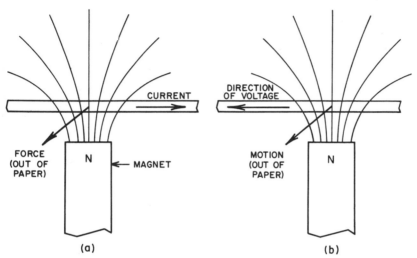

FIG. 4. (a) Force on a current-carrying wire lying in the magnetic field of a magnet. (b) Voltage produced in a wire moving through a magnetic field.

ent time the magnetic field around a magnet is known to be caused by electric currents circulating around inside small regions in the magnet, that go on circulating indefinitely. In an unmagnetized piece of iron, these currents circulate in random directions, and so essentially cancel one another. Under the influence of another magnetic field, however, the currents may line up to all circulate in the same direction; the material then becomes magnetic and will itself produce a magnetic field. The forces that magnets exert on each other are thus basically due to the forces that the currents in the magnets exert on each other.

To sum up: A wire carrying an electric current experiences a force if it is lying in the magnetic field of a magnet or another current; so if it can move, it will do so. This has important practical applications that we will indicate shortly.

Another effect exists that is somewhat the opposite: A magnetic field can produce a current. If a wire moves in a magnetic field, or if the magnetic field changes in magnitude while the wire lies in it, a voltage is produced in the wire, which may cause a current to flow. In Fig. 4(b), for example, if the wire is moved in a direction out of the paper, a voltage appears that can produce a current in the direction of the arrow. This effect is the basis of the generation of electric voltage and current by mechanical means, and is also of great practical importance.

These two effects are the physical basis for a class of devices called *transducers,* which can transform mechanical forces and motions into electrical currents and voltages and conversely. The voltage or current produced by a transducer is frequently called a *signal.* Fig.. 5 is a cross-sectional representation of one such device. A coil of wire is wound on a short cylinder that is placed in a cylindrical gap between two pieces of iron of the general shape shown in the figure. These are

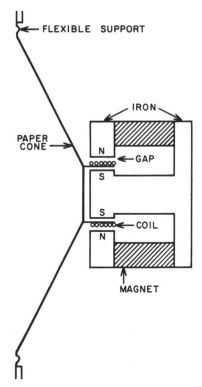

FIG. 5. A transducer which can function either as a dynamic microphone or as a dynamic loudspeaker.

joined to a piece of magnetized material as shown. The iron becomes magnetized by being in contact with the magnet and produces a magnetic field in the cylindrical gap. If a current is made to flow in the coil of wire lying in this magnetic field, the coil experiences a force to the right or to the left depending on the direction of the current. If this coil is attached to a cone of some reasonably rigid material held at the outer edge as shown by a flexible support, the force on the coil will pull the cone in or out. If the current flowing through the coil alternates in direction, the cone will vibrate back and forth and can radiate sound. This system comprises one form of *loudspeaker;* one constructed according to this scheme is called a *dynamic* loudspeaker.

This system will also work in reverse. Sound waves striking the cone will push it back and forth and therefore move the coil across the magnetic field in the gap. This in turn will generate a voltage in the coil, which will appear across the terminals of the coil. Devices such as this that can convert sound waves into electrical voltages or currents are called *microphones;* this particular style is called a dynamic microphone.

Microphones comprise a very important class of transducers, and there are other types that do not use the principle outlined above. One rather inexpensive type utilizes the peculiar property of certain crystalline and ceramic substances to develop an electrical voltage when they are pushed out of shape; these, naturally enough, are called *crystal* or *ceramic* microphones. Another type, called a *condenser* microphone, is used when high-quality transformation of sound signal to the electrical signal is wanted. This type is sketched in Fig. 6. It has a stretched con-

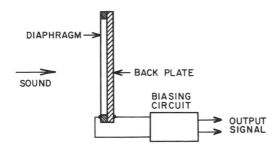

Fig. 6. Cross-section of condenser microphone.

ducting diaphragm placed close to a conducting back plate, with both connected to a circuit that maintains an electrical voltage between them. Sound waves striking this diaphragm cause it to oscillate and change the spacing between it and the back plate. This causes a corresponding change in the voltage between the diaphragm and the plate.

Two types of transducers useful for obtaining electrical signals from vibrating systems are shown in Fig. 7. That shown in Fig. 7(a) is an application of the condenser microphone principle. A conducting plate is placed near the vibrating object, which must also be a conductor.

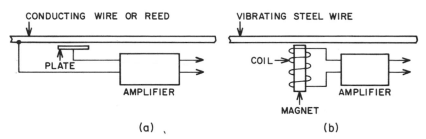

FIG. 7. Transducers for converting mechanical vibrations of a wire or reed to electrical oscillations. (a) Electrical method. (b) Magnetic method.

Both are connected to an electrical circuit which produces a voltage output that changes with the spacing between the two. That shown in Fig. 7(b) is an adaptation of the dynamic microphone principle. The vibrating object must be magnetic; as it moves, it changes the spacing between itself and the nearby magnet which has a coil of wire wound on it. This changes the magnetic field in the coil and produces a voltage in it.

All of these devices have limitations of various sorts. In particular, the microphones develop quite small voltages for ordinary sound pressures; conversely, the loudspeakers require fairly large currents in order to develop reasonable sound pressures. Used by themselves, the transducers would not be of much practical use; they are made so by being used in connection with an *amplifier,* an electrical circuit that can transform small voltages and currents into much larger ones. The heart of the amplifier is a device called a *vacuum tube,* in which the flow of a relatively large current can be controlled by a small voltage on one of its electrical terminals, just as a large flow of water can be controlled by a small force on the handle of a valve in the pipe. A single vacuum tube can develop an amplified voltage (called the *output*) that is as much as 50 to 100 times the voltage applied to it (called the *input*). By using the output of one tube as the input of another, considerable over-all amplification of voltage can be obtained.

In recent years devices called *transistors* have been developed. These function in somewhat the same way as vacuum tubes, a small current put into the device resulting in a much larger current coming out. Transistors have a number of advantages over vacuum tubes and are gradually displacing them in electronic circuits.

By the use of amplifiers, the small signal produced by a microphone in a sound of low intensity can be built up as much as desired. This amplified output can then be supplied to a loudspeaker and converted back to sound of much greater intensity than the original.

Electronic Recording and Reproduction of Sound

The first applications of electronics were to the recording process, making it much easier and more flexible. During recording sessions, the musicians no longer had to sit as close as possible to the recorder horn and play as loudly as they could; instead they could sit and play normally, with strategically placed microphones picking up the sound. The outputs of these microphones could then be mixed as desired; for example, one section of an orchestra could be emphasized, or a soloist could be made to stand out more loudly above the rest of the ensemble. The resultant mixture of microphone outputs, properly amplified, would be fed to another transducer to operate the stylus cutting the wax master record. The problems caused by mechanical resonances in the horn and diaphragm system were completely eliminated. Other problems of an electrical nature inevitably replaced them, but these are generally easier to deal with.

The reproduction of sound from records likewise benefited. Instead of using the displacement of the grooves in the record to generate sound directly, they are used to move a stylus on a transducer to generate a small electrical signal. This signal is then amplified and fed into a loudspeaker, from which as much sound as is wanted may be obtained. As a consequence, the size of the stylus and the groove for electrical reproduction may be considerably smaller than that necessary for mechanical reproduction. Earlier recordings were made with a record rotation of 78 revolutions per minute and with grooves of such width that 4 to 5 minutes of playing time could be obtained on one side of a 12-inch record. Advances in transducer design have made it possible to use a rotation speed of $33\frac{1}{3}$ revolutions per minute together with considerably finer grooves, so that at present more than 20 minutes of music may be recorded on one side of a 12-inch record. The present-day transducer, called a *cartridge* or *pickup*, uses a diamond stylus, which is very hard and can be used a long time without excessive wear. It is mounted in such a way as to exert a very small force on the record, keeping record wear to a minimum.

Tape Recording

We mentioned above that some substances have the property of becoming magnets themselves when placed in the field of another magnet, and that in some cases this acquired magnetism may be retained when the substance is removed from the field. This is the basis of a recording process by which an electrical signal produces a varying amount of magnetization in a long strip of such magnetic material. The idea is not new, for the process of recording on a fine steel wire was first worked out about fifty years ago. The commercial and practical application of magnetic recording has come about more recently as a result of the development of magnetic tapes, hence the term *tape recording*.

The process of tape recording is illustrated in Fig. 8. The tape consists of a narrow strip of thin plastic ribbon, coated on one side with a

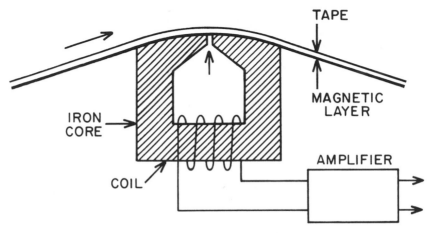

FIG. 8. Tape recording and reproducing head.

very thin layer of iron oxides; these oxides are among the substances that can be permanently magnetized. This tape is then drawn at a constant speed across what is called a *recording head,* which consists of an iron core shaped somewhat as in Fig. 8, having a coil of wire wound around one portion and having in it a very narrow gap at the point where the tape is in contact. A current flowing in the coil of wire produces a magnetic field across the narrow gap, some of which penetrates the thin layer of iron oxide on the tape. As the tape moves across the recording head, the amount of magnetization produced in the magnetic layer varies with the amount of current flowing in the coil of wire on the head. Variations in the electric current in the recording head are then recorded as variations in the magnetization along the length of the tape.

This process may now be reversed. If the tape is rewound and then moved across the head in the original direction, the variations in magnetization of the tape produce variations in the magnetic field in the wire coil; these in turn produce a varying voltage in this coil. The iron core with its coil is now serving as a *reproducing head.*

The current in the recording head can be obtained from the amplified output of microphones picking up sounds to be recorded on the tape. Subsequently, the voltage from the reproducing head may be amplified and fed to a loudspeaker to re-create the original sound. With proper construction of the equipment, proper design of amplifiers, and so on, it is possible to get quite a faithful reproduction of the original sound.

The upper limit to the frequency that can be recorded on tape depends on the width of the gap and the speed of the tape. If an alternating current flows in the coil of the recorder head, it produces an alternating magnetization in the tape moving across the head, providing it is moving fast enough. However, if the tape is moving so slowly or the gap in the head is so wide that the tape does not move very far across the gap during one alternation of the current, the magnetizing effect on the tape is essentially cancelled out. A signal of this frequency is then not recorded as well as lower frequencies, and there will be a frequency above which signals will not be recorded at all. A tape speed in common use today is 7½ inches per second, fast enough to give reasonably good high-frequency reproduction for musical purposes. If high-quality reproduction is not necessary, as in the recording of speech sounds where high frequencies are not needed, slower speeds may be used; small battery-operated tape recorders are available which use half the speed (3¾ inches per second) or one-quarter the speed (1⅞ inches per second).

The same head may be used for both recording and reproducing, as outlined above, but the present-day practice in good machines is to use separate heads. In addition, the tape should be completely unmagnetized when it is used for recording. This is accomplished by using a third head, ahead of the other two, that has a high frequency alternating current flowing in it. This removes any previous magnetization that may have been on the tape.

An unmagnetized tape moving across a reproducing head actually produces a small randomly fluctuating voltage in the coil of the head. This is because the particles of iron oxide on the tape are not actually unmagnetized; they are magnetized but pointing in random directions. At any instant, as the tape passes across the head, the magnetic fields produced by the small particles of oxides average out almost to zero but not quite, because at any one instant there may be a few more magnetized particles pointing in one direction than another. The randomly varying voltage produced in the head constitutes noise, such as we discussed in Ch. 6. It appears in the output of the tape recorder as a "rushing" or "hissing" sound. This noise background puts a limit on how small a signal may be recorded, since one too small will be masked by the noise.

There is also an upper limit on the strength of signal that may be recorded, because the iron oxides can be magnetized only so far, at which point they become *saturated* and refuse to retain more magnetism no matter how strong the current in the recording head. This effect will produce distortion, so that the reproduced signal is no longer an accurate copy of the recorded one. In a good tape recorder, the range between amplitudes of the loudest and softest sounds that can be practically recorded is about 100:1, or 40 decibels.

An interesting variation of the tape recorder has recently been developed.[1] This machine uses several reproducing heads mounted on a structure that will rotate as the tape is passed over it, but in such a way that at any time one head only is in contact with the tape. When one head leaves the tape, another takes its place. The heads will be moving with or against the tape depending on the direction of rotation of the structure supporting the heads. If a head is moving with the tape, the reproduced frequency of a signal recorded on the tape will be lowered from its actual value by the Doppler effect described in Ch. 3. Conversely, if the head is moving in the opposite direction to the tape, the reproduced frequency will be higher. If a tape having recorded music is played on this device, the pitch of the reproduced music may be raised or lowered considerably by the direction and rate of rotation of the heads. On the other hand, the tempo of the music is set by the speed of the tape across the head system, and is not changed. Alternatively, the tape speed may be changed in such a way that the pitch of the music is not altered but the tempo is varied. On the usual record player or tape recorder, both the pitch and tempo of recorded music will change if the record or tape speed is changed, but they both change together and in the same way. It is quite striking to hear them being changed independently, as is possible with this machine.

High-Fidelity Sound Reproduction

The development and perfection of the various transducers and associated devices has resulted in a whole industry devoted to the manufacture of "hi-fi" equipment. The recording of sound on phonograph records has reached a high state of perfection.[2] The pickups used to reproduce the recorded sounds are available in a variety of constructions and prices. The pickup is mounted on the *tone arm*, which allows the stylus to ride gently in the record groove and move freely across the surface. The tone arm is mounted adjacent to the *turntable* which rotates the record. The mechanical and electrical design of this equipment is the result of a great deal of research and development by the manufacturers. This is true also of amplifiers, available in many sizes and designs and with large power outputs. Many excellent speaker systems have been designed and the cost of this particular component of a high-fidelity system may be quite considerable.

Equipment of this sort is susceptible to troubles of various kinds, which the manufacturers try to keep within reasonable bounds. Some of them are described by terms of a rather picturesque character. The turntable which rotates the phonograph record may produce *rumble,* which is caused when low-frequency vibrations from the turntable motor are transmitted to the record pickup. Slow variations in the turntable motor speed appear as a corresponding variation in pitch of the reproduced sound called *wow;* this may also be caused by the hole in the record being off center. A more rapid variation in turntable speed appears as a pitch fluctuation called *flutter.* Other problems of a mechanical nature may be present. The tone arm carrying the pickup may have a resonance which will accentuate low frequencies, and resonances may occur in the pickup itself that accentuate high frequencies.

There are also difficulties of an electrical nature. The amplifier should have a *flat* frequency response; that is, it should amplify signals of all frequencies by the same amount. (However, circuits called *tone controls* are usually built into an amplifier so that high or low frequencies can be accentuated or attenuated as desired.) The signal coming out of an amplifier should be a faithful reproduction of that fed into it; a sinusoidal signal input should appear as a sinusoidal output. If this is not so, harmonics are introduced into the output of the amplifier that are not present in the original signal. This is called *second-harmonic distortion,* since this harmonic is generally the most prominent of those the amplifier produces. Another trouble related to this, termed *intermodulation distortion,* may be present; it results in an input signal containing two frequencies giving an output that has in it not only these two frequencies, but also frequencies that are the sum and the difference of the two frequencies.

Loudspeakers may introduce distortions by accentuating certain frequencies over others.[3] It is difficult to construct a single loudspeaker that will reproduce all frequencies in the audio range equally well, so common practice is to use one loudspeaker for low frequencies and another one for high frequencies, named the *woofer* and *tweeter* respectively. The output signal from the amplifier is fed to these loudspeakers through what is called a *cross-over network,* which splits a signal into its low-frequency and high-frequency parts and sends each part to the proper loudspeaker.

In earlier sound-reproducing systems the high-frequency range was quite restricted; for a considerable period of time all frequencies above 5000 cycles per second were not reproduced. This went on for long enough to condition listeners quite well to this restricted frequency range; thus, as equipment was improved and the upper-frequency limit raised, a certain amount of listener re-education was necessary to get acceptance of the wider and more natural range.[4] At the present time, a well-designed sound system will reproduce frequencies to well past the upper limit of hearing—past 15,000 cycles per second.

The development of tape recording added still more flexibility to the recording process. In the earlier days of record making, one entire side of a record, amounting to 4–5 minutes of music, had to be made at one time. Alterations in the wax master record were not possible, and if any mistakes occurred in the playing of the music, the whole record had to be made over. The advent of long-playing records would have aggravated this problem considerably except for the fortunate circumstance that the development of tape recording came along just in time to alleviate it. At the present time the original recordings of musical performances are made on tapes rather than wax masters. Tapes may be cut and spliced together again in any desired way, so that it is possible to record on tapes several performances of the same work and then combine one part of one performance with another part of a different performance. In this way not only may mistakes be eliminated, but those parts of each performance judged best may be combined into a single tape, from which disc recordings may subsequently be made. The final result will then be a product that has a perfection not present in any one individual performance. Some ethical arguments might be pertinent to this procedure; it does seem that the listener should be informed of this practice so he will not be conditioned to expect from live performances the technical perfection of recorded ones.

Stereophonic Sound Reproduction

A recent development is the commercial production of "stereo" equipment to give added realism to sound reproduction. The idea of *stereophonic* sound reproduction is not new—it goes back about one hundred years, and its further development as applied to sound recordings was worked out some thirty years ago—it is only the commercial development that is recent.[5]

We are accustomed to listening to sounds with two ears. The sound signals arriving at the ears will differ somewhat in intensity, time of arrival, and phase, and the brain sorts out these differences in such a way as to enable us to tell the direction from which the sound comes. If we listen to sound coming from a single loudspeaker, it will obviously be coming from the direction of the loudspeaker, regardless of the location of the microphone that picked up the original sound. However, if we listen to sound coming from two loudspeakers placed reasonably far apart, and if these are reproducing sounds from two microphones placed near the sound source and having a reasonable separation, our ears will interpret the sound from the two loudspeakers as coming from a direction that depends on the position of the original sound source in relation to the two microphones. If we place the two microphones in the vicinity of an orchestra, record their outputs separately, and then reproduce these through two separate loudspeakers, the sounds produced by the instruments in the orchestra will appear to come from different directions, and an illusion of dimensionality will be added to the reproduced sound. Stereophonic sound reproduction thus requires two completely separate systems or *channels* from each microphone to each speaker.

Stereo tape recording is accomplished by using two separate recording heads, each recording on half the width of the tape. The so-called *four-track* recording uses the same principle, except that the recording heads are narrow enough so that the two of them use only one-half the total width of the tape; by turning the tape over and running it the other way, we use the other half of the tape and obtain twice as much recording time.

The two channels necessary for stereo recording on records are obtained in a somewhat more complicated manner, as we see in Fig. 9(a). For so-called *monophonic* (single-channel) recording, the record

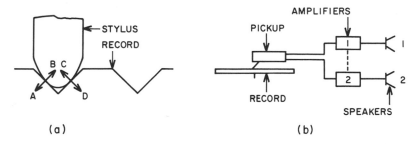

FIG. 9. Two-channel system for stereophonic sound recording. (a) Obtaining two channels on the record. (b) Two channels from pickup to loudspeaker.

groove is the same depth everywhere but is displaced from side to side in accordance with the sound signal. For stereophonic recording the two sides of the groove are inclined at 45-degree angles as shown; the signal from one channel moves one side of the groove in the direction indicated by the arrow and the other channel moves the other side. The pickup is so constructed that two separate electrical outputs are obtained, one corresponding to the displacement of one side of the groove and the other output corresponding to the other side. These two signals are then amplified and fed to the respective loudspeakers. Fig. 9(b) shows a sketch of such a system.

The stereo effect can sometimes be so great that the orchestral instruments seem to be grouped some on one side and some on the other, with nothing in between. For this reason, the amplifiers of the two channels are often interconnected as shown by the dotted line in Fig. 9(b), so that some of the output of one channel may be mixed with the other, and conversely; this alleviates this "hole-in-the-middle" effect.

It is important in setting up a stereophonic system that the proper channel be connected to the proper loudspeaker, and not reversed, otherwise the orchestra will appear to be turned around. It is also important that the loudspeakers be "phased" properly, so that the loudspeaker cones move in the same direction rather than opposite directions when the two sides of the record groove move together. If this condition is not met, the stereophonic effect will be seriously degraded.

Generation of Sound by Electronic Means

The electrical oscillations that can be used to produce sounds do not have to be made by other sounds; they may be made directly in the electrical circuits themselves. The idea of generating sounds from electrical oscillations is not new (it goes back to before the turn of the century), but again it is the technological development of electronics that has made possible the practical application of this idea. The electrical signals we want to turn into sound may be quite small so amplification is generally necessary; hence the need for electronics.

Electrical oscillations to produce sound may be obtained in a great variety of ways. One relatively straightforward way is to use a vibrating system to produce the electrical signals instead of producing sound directly. Since these signals may be amplified as much as desired, the sound finally resulting may be very much greater than that which the vibrating system could supply directly.

An early application of this idea was an electronic piano. If the soundboard of a piano is removed, the vibrating strings are practically inaudible. However, by placing a transducer of the type shown in Fig. 7(a) or (b) somewhere along the string, the vibrations may be used to generate electrical signals, which can then be amplified and fed to a loudspeaker. The quality of the sound produced can be varied somewhat by varying the position of the transducer along the string;

for example, if it is placed at a point that is a node for a particular mode of oscillation of the string, that particular mode would not generate any sound output. With no soundboard to carry energy away from the strings, their vibrations can last a good deal longer; this gives the sound of this kind of piano a somewhat organ-like quality as compared to the conventional piano. This particular instrument was in commercial production for a time, but has now become obsolete.

Another electronic piano, commercially available at the present time, is based on the same principle, but uses vibrating metal reeds instead of strings. These reeds are struck by hammers which are operated by the conventional piano mechanism; the reed vibrations then generate electrical signals which are amplified and reproduced. The reeds are so constructed that their vibrations die away at about the same rate as the string vibrations of a conventional piano. However, since the vibrations of a metal reed have partials quite different from those of a string, the sounds of this piano do not at all resemble those of a conventional piano. It is also provided with an electronic vibrato, which can be switched on to make it sound even less like a normal piano.

The old-fashioned reed organ used metal reeds of the sort described in Ch. 11 in connection with the pipe organ. Air under pressure was supplied by bellows pumped with the feet and blown through a given reed by pressing the appropriate key on the organ manual. This organ was also subjected to electronic treatment. The reeds were isolated by sound-insulating materials so that their vibrations could not be heard. They were then all set into vibration simultaneously by air obtained from a blower. The individual reed vibrations were then picked up by individual transducers of the sort shown in Fig. 7. Pressing a key on the organ manual would then connect the appropriate transducer to the amplifier and speaker to produce an amplified and altered sound of the reed. An instrument of this type was commercially available for a time, but has also become obsolete.

The electric guitar, on the other hand, has become very popular. In the conventional guitar, as in the string instruments, the vibrations of the strings are transmitted to the body of the instrument and then radiated into the air. In an electric guitar, the sound radiated by the body is insignificant; instead, the string vibrations are picked up by a transducer located on the bridge and are then amplified and reproduced. In this way very high sound levels may be obtained, which can actually be injurious to the ears of the listener, as mentioned in Ch. 5.

Recently, similar devices have been developed to attach to violins, clarinets, and so on, so that these instruments can compete with the electric guitar. Not only will such devices provide these instruments with an astonishing sound output, but the sound may be altered electronically to a degree limited only by the ingenuity of circuit designers. For example, a circuit called a *ring modulator* has been designed, which, at the start of each individual tone fed into it, can produce rapidly decaying electrical oscillation of half the frequency.[6] By using this circuit together with an amplifier and speaker, a singer may accompany himself with a simulated pizzicato accompaniment pitched an octave lower.

Electronic Organs

At the present time the most extensive application of electronics to the direct production of sound (rather than sound reproduction) is the *electronic organ*. This device uses circuits which generate electrical oscillations directly; these are then converted into sound.

One of the early versions of the electronic organ, still in production, is based on the generation of electrical oscillations by the means illustrated in Fig. 10. The magnetic transducer illustrated in Fig. 7(b) is

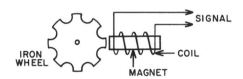

Fig. 10. Magnetic method of generating electrical oscillations.

placed close to a rotating iron wheel whose periphery contains serrations or projections as shown. As the wheel is rotated, the periodic change in the spacing between the magnet and the wheel produces a corresponding voltage in the transducer coil. By properly shaping the wheel periphery and by putting the electrical output of the coil through appropriate circuits, a practically sinusoidal electrical output may be obtained. By mounting on a single shaft wheels having 2, 4, 8, 16, and so on projections around the periphery, tones an octave apart will result. With twelve such groups of wheels mounted on shafts geared together with gears of the proper ratio (as in the stroboscopic frequency meter described in Ch. 8), tones may be obtained whose frequencies are those of the tempered musical scale.

Since this device produces pure tones, complex tones are obtained by an additive process. A given tone, as fundamental, has added to it the octave above, the twelfth above, and so on, the amounts of each being proportioned by switches connected to tabs that are pressed to obtain the particular mixture wanted. The partials of the tone produced in this fashion cannot be called harmonic in the strict sense; for example, the fifth partial would be added by using the tone two octaves and a major third above the lowest tone, and since this third is in the tempered scale its frequency will not be exactly five times the fundamental frequency. Furthermore, seventh harmonics will be missing, since there are no notes in the tempered scale close enough to be used. However, these deficiencies do not seem important, since this organ is very popular.

Other models of electronic organs use circuits that produce electric oscillations directly without using wheels or other moving parts. A circuit that produces a sinusoidal electrical oscillation whose frequency is in the audible range is called an *audio oscillator*. It is easy to construct electronic circuits that will oscillate; in fact, it is sometimes hard to prevent them from doing so. The difficulty lies more in designing circuits that will maintain a sufficiently constant frequency regardless of aging of circuit components, changes in power line voltage, and so forth.

One of the first musical applications of this idea was the "Theremin," an electronic oscillator generating a single tone whose frequency could be varied by moving one hand closer to or farther away from a metal rod; the loudness was similarly controlled by moving the other hand in the vicinity of a metal loop. Another application was a device generating a single tone whose frequency could be varied by means of a keyboard; it was designed to be attached to a piano to provide an additional solo voice. These single-tone devices are for the most part obsolete, except for their use in experimental work on the composition of musical sounds.

Presently available electronic organs utilize circuits that produce complex waves rich in harmonics, such as the square waves and sawtooth waves discussed in Ch. 6. Common practice at the present time is to use twelve oscillators tuned to the frequencies of the tempered scale and occupying the top octave of the range to be covered, such as $C\sharp_7$ to C_8. Notes in the lower octaves are produced by electronic frequency division; it is relatively easy to construct a circuit that will deliver a frequency exactly half as great as that fed into it. With a number of such circuits connected in series, a single oscillator in the top octave producing, say, $C\sharp_7$ can generate the $C\sharp$ for all octaves below. The output of the whole system is then a series of complex electrical oscillations covering every frequency in the tempered scale. Pressing a key on the organ manual selects the proper frequency and puts it into the amplifying circuit.

To vary the quality of the tone produced by the oscillator, a subtractive process is used. The signal is put through electrical circuits called *filters*; these allow certain frequency ranges to pass through them and reject others. This changes the harmonic content of the original electrical oscillation and so alters the quality of the tone it produces. By using differently proportioned filters, quite a range of qualities may be achieved, the appropriate filters being selected by tabs placed above the organ manual. Since this system works in much the same fashion as the human voice, whose quality depends on the formant regions, these filters are sometimes called formant filters.

Both the above types of organs—the additive and the subtractive—suffer from a major drawback as compared to a conventional pipe organ. For a given stop on the conventional organ, playing a note an octave above a given note adds another pipe to the system, and so provides another sound source. The ear hears the two pipes as producing two separate notes, since the higher pipe will never be exactly an octave above the lower. In the electronic organ, on the other hand, the frequency relationships are exact. Adding a note an octave higher is then equivalent to changing the amplitudes of the even harmonics of the given note, so the effect is of a single note with an altered quality. Similarly, in the conventional organ, adding stops adds new ranks of pipes, which are all separate sound sources of very nearly but not quite identical frequencies. The tone is then enriched by the chorus effect that we mentioned in Ch. 6. In the electronic organ, however, adding stops only changes the harmonic structure of a single complex tone, since all tones come originally from the same oscillators, and the chorus effect is lacking.

To obtain this chorus effect and get a sound more like that of a conventional pipe organ, more elaborate electronic organs are now being built with one or more electrical oscillators for each note in the keyboard. This arrangement avoids the difficulties connected with the mathematical exactness of the simpler models, but is obviously more expensive.

The sound of an organ in the usual home living room will lack the reverberation associated with organ sound in a church or large hall. To add this effect, various artificial reverberation devices have been developed. One method in practical use is to convert the electrical signal to a mechanical motion by means of a transducer connected to coiled spring; the disturbance produced travels along the spring and arrives at a transducer at the other end at a later time, where it is converted back to an electrical signal.[7] By the use of a number of springs with properly proportioned lengths, a rather realistic reverberation effect can be produced. Another method uses a rotating loop of magnetic tape; the electrical signal is recorded on the tape at one point and taken off by reproducing heads placed at various points along the tape so that successive signals are obtained, giving the effect of reverberation.

Vibrato effects are also supplied. Electronically, it is easy to produce a periodic change in the amplitude of an electrical oscillation; however, this is not a true vibrato, since it does not have the periodic frequency change necessary for proper vibrato quality. In some systems this periodic frequency change is produced electronically in the oscillating circuits themselves. Other mechanical arrangements have been devised utilizing rotating vanes, rotating speakers, and so on, that provide the periodic frequency change by the Doppler effect.

It is not difficult to construct electrical circuits in which an oscillation, once started, dies out more or less rapidly. Percussive effects may thus be obtained; for example, a relatively pure tone that dies out slowly will resemble a bell, and one that dies out rapidly will sound like a xylophone. A great deal of ingenuity has been used in devising such effects.[8]

Further and more detailed information on electronic organs is available in the literature.[9]

Synthesizers of Musical Sounds

A logical extension of the electronic organ is one that will make its own music rather than utilizing a performer; such an arrangement is then a *music synthesizer*. One such device has been constructed by an industrial organization prominent in the communications industry.[10] This instrument is basically an elaborate electronic organ played from a roll of perforated paper, as was done with the old-fashioned player pianos.

The basic electrical oscillations are produced by a set of fixed frequency oscillators giving notes from C_9 to B_9, plus two sets of variable frequency oscillators that can be tuned to these frequencies or to any other desired frequencies. The outputs of these oscillators go to a frequency dividing system so that all of the notes of the musical scale from C_0 to B_9 may be obtained. A long roll of paper has holes punched in it at appropriate places; these allow electrical contacts to be made that select the frequency of the tone and also control the various electrical circuits determining the quality, loudness, and so forth of the tone. For example, eight spaces along the width of the paper are allotted to determining the pitch of the tone; four spaces determine the octave, and the other four spaces determine the note within the octave. The system used is called *binary coding* and it is quite economical, providing sixteen different choices by using four spaces in the manner shown in Fig. 11. Since the paper travels at a constant speed, the duration of the tone is determined by how many holes are punched along the length of the paper. Other spaces along the width of the roll are used to determine the loudness of the tone and its envelope. Remaining spaces on the roll are used to specify the quality of the tone by switching in and out formant filters in order to remove varying amounts

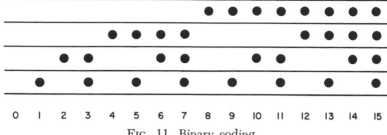

0 1 2 3 4 5 6 7 8 9 10 11 12 13 14 15

Fig. 11. Binary coding.

of harmonics from the sawtooth waves produced by the oscillators. Four separate musical lines, each controlled in this way, can be produced simultaneously, and recorded on tape synchronized with the paper roll. If desired, four more tones can be produced from another roll of paper and recorded on another band of the same tape in synchronism with the first four tones. With four-track tape, sixteen simultaneous musical lines can be recorded. The outputs of the various bands on this tape may then be reproduced, treated further if desired —by adding reverberation, for example—and finally combined on a single tape.

This machine enables the composer to produce music without having to use the services of a performer. Instead of writing notes on paper, the composer punches holes in the roll of paper by means of a machine constructed for the purpose and specifies the frequency, loudness, quality, and so on of each of the four notes available. The roll of paper may then be put into the machine and the musical result heard immediately, thus dispensing with musicians, performers, rehearsal time, and many other difficulties connected with making music.

Music from Manipulated Tapes: Electronic Music

The ease with which sounds may be altered and rearranged by electronic circuits and by cutting and splicing magnetic tapes has led to the development of that branch of musical composition originally called "musique concrète," but now described by the general term *electronic music*. Natural sounds, such as musical instruments, singing, speech, street noises—in fact, anything—are first recorded on tapes. These may then be copied at the same speed, or at higher or lower speeds so as to change the pitch of the sounds. Frequencies may be removed by filtering, or reverberation or noise added while the tapes are being copied. Separate tapes may be combined into a single tape by recopying further. Fragments may be cut from tapes and spliced with others to produce any desired succession of these sounds. The final result of all this manipulation will be a tape bringing the conglomerate sounds directly to the listener from the composer without the intermediary of the performer.[11]

The process has its limitations; every time a sound is copied from one tape to another, it loses some of its quality and has a little more noise added to it, so that copying too many times will produce excessive degradation of the original sound.

The composer of this kind of music does not have to confine himself to natural sounds; various electronically produced waves are available, such as square waves, sawtooth waves, sinusoidal waves, and so forth, which can be combined on tapes in this way. A popular kind of sound for this work is that produced by *filtered noise*; this is white noise (containing all frequencies) put through a filter that passes only a relatively narrow band of frequencies, resulting in a noise with a rather definite pitch.

Other methods can be used to add sounds. For example, the individual tones produced by a single oscillator can be combined into chords of any frequency spacing. This is done by first recording the individual tones successively on tape at the desired frequencies, with durations of, say, one second each, and then playing this tape back in a reverberation room. If the reverberation time of this room is long enough, the individual tones will last long enough to form chords, which can then be recorded on another tape.

Considerable experimentation is going on in the field of electronic music, with a number of establishments active.[12] Quite a bit of equipment can be utilized for such work, requiring a considerable financial outlay. A well-stocked electronic music studio will have equipment such as the following:

To supply electrical signals: sinusoidal wave generators; sawtooth and square wave generators; pulse generators; noise generators; devices to generate arbitrary waveforms; microphones; electronic organs.

To modify electrical signals: variable bandpass filters; formant filters; ring modulators; amplitude modulators; frequency modulators; exponential time envelope generators; gating circuits.

To record electrical signals for reproduction as sound: variable speed tape recorders; tape loop mechanisms; tape editing facilities; phonograph; reverberation equipment.

Operating equipment: control panels; patch panels for connecting components; amplifiers, power supplies, monitoring phones and loudspeakers; meters; oscilloscopes; sound level meters; harmonic wave analyzers.

Music constructed with such equipment is available on records.[13]

For a time, a journal was published devoted to articles describing work in the field of electronic music.[14] Unfortunately, much of the language of this journal was in a pseudotechnical jargon that had no real technical basis and whose purpose was apparently to overawe the reader rather than to inform him.[15]

Computer Music

Of the technological advances of recent years, one having a considerable effect on our economic organization is the development of the electronic computers. Mechanical machines for doing the arithmetical operations of addition, multiplication, and so on have been around for a long time, but in recent years electronic circuits have been devised that can do the same operations and at a much greater speed. Such an electronic calculator is called a *computer*. Not only can the computer calculate with great speed; it can also be provided with a *memory* in which it can store the results of its calculations for later use. The computer performs its arithmetical operations in accordance with instructions provided to it by some means, such as punched cards; a set of instructions is called a *program*. Instructions may also be stored in the memory for later use. Computers are doing more and more of the drudgery associated with our complex economic structure, and it is natural for musicians to wonder what applications there might be for computers in the field of music.

One important application has already appeared. The computer is very useful in determining what kind of sound will be produced by some arbitrarily specified waveform. For example, we might want to know the sound produced by combining two pure tones of arbitrarily selected frequencies, such as, say, 500 and 610 cycles per second. The computer can be instructed to calculate the result of adding two waves of these frequencies at specified amplitudes at intervals of, say, 0.0001 second. The results of its calculations will be a series of numbers representing the amplitude of the resulting wave for each of these intervals; it will appear as a series of numbers (in binary form, as illustrated in Fig. 11) recorded on a magnetic tape. This tape can then be put into a device called a digital-to-analog converter, which transforms the numbers on the tape into electrical signals proportional to the size of the numbers. These signals can be recorded on magnetic tape and then played back in the usual way, so that the sound resulting from adding these two frequencies can be heard.

The example we started with is a rather simple one and could be carried out with a couple of tuning forks; but in principle any arbitrarily complicated waveform can be heard in this way. Investigations of this sort with a computer have been very useful in determining what aspects of musical tones are important. For example, the tone produced by an electrical oscillator of constant frequency and amplitude is quite dull and uninteresting, and becomes irritating if listened to long enough. However, such a tone acquires added richness and interest if its amplitude is varied over a range of about 50 percent and its frequency over a range about 1 percent in a random way at some eight to twenty times per second.[16] A tone with this kind of variation is just about what a good human performer would produce; actually, a good deal of practice is necessary to produce tones on an instrument that do not vary a good deal more than this.

By analyzing the tones of musical instruments and then synthesizing them again on a computer, it is possible to determine what aspects of a musical tone are important by altering the synthesized tone in various ways. In this manner it has been found, for example, that to specify a tone only by its harmonic structure is quite inadequate.[17]

Theoretically, the computer could be used to generate any desired musical output; however, to get reasonable quality, we would need to calculate some ten thousand numbers per second for the duration of the musical composition. This is rather impractical, even if we knew how to specify in advance the waveforms necessary to get the desired sounds; this we are not yet able to do. For this reason somewhat more limited but more practical schemes have been worked out to utilize the computer.

One such method is to first store in the computer memory a series of numbers representing the amplitude of a particular waveform at successive parts of its cycle. A number of such waveforms may be stored simultaneously. The computer may then be instructed to reproduce one of these waveforms a prescribed number of times per second and at a prescribed amplitude. It is then functioning as an "instrument" whose frequency and loudness can be specified, although its output is in the form of numbers rather than sound. The "instrument" is "played" by means of instructions to the computer in the form of punched cards specifying the frequency, intensity, and duration of each note. Several such "instruments" may be utilized simultaneously and their outputs added. The final output, representing the "ensemble" of these "instruments," will again be a series of numbers on a magnetic tape; it may then be converted to sound by using the digital-to-analog converter as described above.[18]

One of the present problems in the use of a computer in this way is the time lag—some hours to days—between the composer's instructions to the computer and the realization of the actual musical output as sound.[19] This is not unusual, however; a composer frequently has to wait some considerable time to hear his work performed. Another problem with the computer is the expense: to produce a few minutes of music may require some ten times as much computer time at a cost of several hundred dollars per hour. Again, this is not unusual; symphony orchestras are also quite expensive. Examples of computer-generated music may be heard on records.[20]

A computer has been adapted to produce music in "real time," which means that instead of waiting for the computer to complete its calculations, the sound output can be heard immediately.[21] The computer can be arranged to produce electrical square waves directly by making electrical connections to its internal circuits; the frequencies of these waves can be specified by instructions to the computer. The square waves are then put through circuits that transform them into sawtooth waves, which in turn go through a circuit, controlled by the computer, that determines their amplitude and envelope. Other sawtooth waves of two and four times the frequency can be similarly produced and controlled and mixed with the original to obtain a "voice" with a quality that can be varied somewhat. This can now be mixed with other "voices," filtered as desired, and then supplied to the usual amplifying circuits and loudspeaker. In this fashion the composer programs in advance the frequency, duration, and so on of the notes he wants, feeds this program into the computer, and then can hear the output immediately. However, the advantage of immediate hearing is balanced by the smaller flexibility of this system; instead of being able to choose from a variety of waveforms stored in the computer, only square and sawtooth waveforms can be used.

Composing by Computer

Since these applications of the computer, producing music directly, do away with the performer, the logical next step is to use the computer to compose music, and so to dispense with the composer. (The logical third step, that of letting the computer also listen to this music, does not appear to have been tried as yet.)

The computer can be instructed to pick out numbers at random; such numbers are called *random integers*. These integers may be correlated with musical aspects such as pitch, intensity, and so on. For example, the notes from C_2 to C_6 could be correlated with the numbers 1–49. The computer selects one of these numbers at random—say, number 15 representing D_3. It then compares this selection with previous ones to see if it satisfies certain rules of counterpoint that are also programmed into the computer. If the note selected satisfies the rules, it is retained; if not, it is rejected, and another one tried. The notes accepted are stored in the computer's memory and are finally punched out on a paper tape; they may then be transcribed into ordinary musical notation.[22] Results of such composition by computer are available on records.[23]

In the composing process outlined, it would be quite simple to change the programming of the computer so that it would follow the rules of twelve-tone composition. It could also easily do inversions, retrograde inversions, and so on; in fact, serial music is admirably suited to the computer, and could be turned out in huge quantities.

Another adaptation of this composing process that has been tried was to have the computer select notes not completely at random, but with a probability that depended on the previous note or pair of notes. By determining the probability values from the analysis of two- and three-note combinations in eleven songs of Stephen Foster, a computer was arranged to compose quantities of what was called "Stephen Foster type" of music.[24]

It is too early to predict with any confidence the future status of the computer in the field of music. It can certainly be of tremendous benefit in such chores as analyzing music, transcribing and extracting parts from scores, and so on.[25] It is not possible to judge as yet the extent to which the computer will replace the performer or composer as an important element in the musical situation. Those who feel that music is an emotional communication between a composer and a listener through the intermediary of a performer will find it difficult to imagine communicating in this fashion with a computer. However, there is a lively interest in the possibilities of electronic and computer music and several centers for such work have been developed and are active.

Research in the Acoustics of Music

An important by-product of the application of technology to the field of acoustics has been the development of specialized equipment for acoustical research. A great deal of equipment is now available to do many tasks that were impossible, difficult, or too time-consuming not too long ago. We have already discussed this briefly in Ch. 5, where we described the early cumbersome and slow mechanical method of analyzing sound waves. Electronic equipment is now available that will analyze a tone into its constituent partials very quickly, making it quite easy to record the spectra of complex sound waves.

In addition to such analyzers, a great variety of instrumentation is now available for research in musical acoustics. There are signal generators that will produce steady or intermittent waveforms of various shapes. There are filters that will pass or stop arbitrary ranges of frequencies. Recorders are available that will record on a paper chart the amplitude of an electrical signal, and level recorders that will record the decibel level of a signal. Meters giving the intensity levels of sounds in various frequency ranges are catalogued by many manufacturers. Numerous transducers can be obtained for converting electrical signals to mechanical forces and motions and conversely. There are many other specialized pieces of equipment available for acoustical work, and more appear all the time, to the dismay of the research worker who is on a limited budget.

Much of the information presented in preceding chapters of this book was obtained by the use of such research equipment, and at the present time we have a fairly respectable body of knowledge in the science of music. However, it should be obvious by now that a great deal must still be learned; in the earlier chapters we were constantly pointing out things we didn't know. Many problems remain to be solved and, since there is good equipment available with which to work, the field of musical acoustics is full of research opportunities.[26]

Unfortunately, it seems that as the opportunities increase and the working facilities improve, the amount of research done in musical acoustics diminishes.[27] The last twenty years have seen a very substantial diminution in the output of work in the field. Part of this is due to the lure of more "fashionable" fields of research than acoustics, particularly musical acoustics. Part of it seems due to a somewhat subconscious attitude prevalent in our society that considers technology an adequate substitute for culture. We hope that this attitude will someday change, and that eventually the attention and financial support devoted to increasing our knowledge of the science of music will be commensurate with its importance.

The musician is generally not very well equipped, either literally or temperamentally, to work in the acoustics of music. However, he should be aware of the benefits that can result from a better understanding of the field. For example, when enough is known about the strings, it should be possible to make really good violins that do not cost a fortune. When we know more about the behavior of the woodwinds, perhaps we can build a bassoon that does not have practically every note out of tune.

One would think that the manufacturers of musical instruments would be the most diligent of those investigating the acoustical behavior of musical instruments. Actually, most of them do not even appear to be particularly interested; they have inherited from their forefathers the methods of making passable instruments, and this seems to be sufficient. This is quite strange for, in every other field of human knowledge that has a basis in science, the application of scientific research has brought tremendous dividends.

Some of these dividends are now in sight, such as better string instruments. More will appear as our knowledge increases.[28] It is our hope that by bringing together in this book the various pieces of knowledge, miscellaneous facts, and scraps of theory that at present make up the accumulated body of knowledge, we will encourage in those concerned with music some appreciation of the value of the science of music, so that they may support its further development.

POWERS OF TEN AND SIMPLE LOGARITHMS

Exponents

In algebra there is defined a notation that is used when a number is multiplied by itself twice, three times, and so on, to form its square, cube, and higher powers. According to this notation:

$$a \times a = a^2,$$
$$a \times a \times a = a^3, \tag{1}$$

and so on. The small superscript number showing the power to which the number a is raised—that is, the number of times a is multiplied by itself—is called the *exponent*.

Scientific Notation

This notation, applied to the number 10 and extended, becomes marvelously useful. We have:

$$10^2 = 10 \times 10 \ = 100,$$
$$10^3 = 10 \times 10^2 = 1000, \tag{2}$$
$$10^4 = 10 \times 10^3 = 10{,}000,$$

and so forth; the number of zeros in the multiplied-out power of ten is equal to the exponent. This is equivalent to writing the number 1, which by implication has a decimal point after it, as 1.00000, and then moving the decimal point a number of spaces to the right equal to the exponent.

The exponents do not have to be positive numbers. From Eqs. (2), we have:

$$10^3 = 10^4/10 = 1000,$$
$$10^2 = 10^3/10 = 100; \tag{3}$$

if we extend this list by following the same rule, we get:

$$
\begin{aligned}
10^1 &= 10^2/10 &= 10, \\
10^0 &= .10/10 &= 1, \\
10^{-1} &= 1/10 &= 0.1, \\
10^{-2} &= 0.1/10 &= 1/100 &= 0.01, \\
10^{-3} &= 0.01/10 &= 1/1000 &= 0.001,
\end{aligned}
\tag{4}
$$

and so on. We see that in this case the value of the negative power is obtained by writing down the number 1. and moving the decimal point to the left a number of spaces equal to the exponent.

This notation is very useful for expressing very large and very small numbers, such as occur often in physics. For example, the distance from the earth to the sun is about 93,000,000 miles. We may write this:

$$93 \times 1,000,000 = 93 \times 10^6 \text{ miles,}$$

or equally well,

$$9.3 \times 10,000,000 = 9.3 \times 10^7 \text{ miles.}$$

As numbers go, this is not a very big one, so not much is gained. However, consider a really large one, such as the number of molecules in a cubic meter of air. This number is 27,000,000,000,000,000,000,000,000, and a great many zeros can be saved by writing it 2.7×10^{25}.

The same notation scheme may be used for very small numbers. For example, by using Eqs. (4), we may write the number 0.0025 as

$$2.5 \times 0.001 = 2.5 \times 10^{-3}.$$

Again, this is not a very small number. For a really small one, we may consider the mass of a molecule of air, which averages about 0.000,000,-000,000,000,000,000,048 kilograms. This can be written much more neatly as 4.8×10^{-26} kilograms.

This method of writing numbers is called *scientific notation,* and it can obviously be a great time-saver. It is customary to put the decimal point after the first digit, as in the last three examples. This notation does not need to be used for numbers of reasonable size; the number of feet in a mile, 5280, could be expressed as 5.280×10^3, but this is hardly worth the trouble. It is for numbers taking more than about four digits that scientific notation is most useful.

This notation has another advantage; it can express the accuracy of a number. If the distance from the earth to the sun is expressed as 9.3×10^7 miles, this means that its value is more than 9.25×10^7 miles and less than 9.35×10^7 miles; this is an accuracy of two *significant figures*. If the distance is measured more accurately, it comes out as 9.29×10^7 miles—more than 9.285×10^7 and less than 9.295×10^7 miles. This is accuracy to three significant figures, and it is ten times better than before, or what is called an *order of magnitude* better. A still more nearly accurate measurement would give the distance as 9.291×10^7 miles, to four significant figures. Zeros to the right of the decimal point are significant. For example, if a given distance on the earth were measured as 200,000 meters, accurate to the nearest meter, this would be written 2.00000×10^5 meters. If it were accurate only to the nearest kilometer, it would be 2.00×10^5 meters.

Numbers expressed in scientific notation are very convenient for calculations. To see why, let us multiply powers of 10. For example:

$$10^2 \times 10^2 = 100 \times 100 \ = \ 10{,}000 = 10^4,$$
$$10^3 \times 10^2 = 1000 \times 100 = 100{,}000 = 10^5,$$

or in general

$$10^a \times 10^b = 10^{a+b}. \tag{5}$$

This gives the general rule: *to multiply powers of 10, add the exponents.*

Similarly, if we divide powers of 10,

$$10^3/10^2 = 1000/100 = 10 \ = 10^1,$$
$$10^2/10^2 = 100/100 \ = 1 \ \ = 10^0,$$
$$10^2/10^3 = 100/1000 = 0.1 = 10^{-1},$$

and so on, or in general

$$10^a/10^b = 10^{a-b}. \tag{6}$$

This gives the general rule: *to divide powers of ten, subtract the power of 10 in the denominator (the divisor) from the power of 10 in the numerator (the dividend).*

With these rules, multiplying and dividing large and small numbers becomes very simple. For example, suppose a rectangle is 1,200,000 feet long and 550,000 feet wide and we want its area. Multiplying in the usual way will produce a surfeit of zeros. In scientific notation we have

$$\text{Area} = 1.2 \times 10^6 \times 5.5 \times 10^5 = 1.2 \times 5.5 \times 10^6 \times 10^5$$
$$= 6.6 \times 10^{11} \text{ sq ft,}$$

since the order of multiplication is immaterial. Similarly for division, picking two numbers at random

$$450{,}000{,}000/150{,}000 = (4.5 \times 10^8)/(1.5 \times 10^5)$$
$$= (4.5/1.5) \times (10^8/10^5)$$
$$= 3.0 \times 10^3,$$

since the order in which the multiplications and divisions are done is immaterial. As one further example,

$$450{,}000{,}000/0.0000030 = (4.5 \times 10^8)/(3.0 \times 10^{-6})$$
$$= (4.5/3.0) \times (10^8/10^{-6})$$
$$= 1.5 \times 10^{14},$$

since to subtract a negative number, we change its sign and add.

Fractional Exponents and Logarithms

How should we interpret a number like $10^{1/2}$? If such fractional exponents are to have any significance, they should follow the same rules as integral exponents. Then

$$10^{1/2} \times 10^{1/2} = 10^{1/2+1/2} = 10^1 = 10,$$

so

$$10^{1/2} = \sqrt{10} \approx 3.16,$$

as we can find from a table of square roots. (The sign $\approx$ means *approximately equal to*; in this case, to three significant figures.) Fractions are rather inconvenient as exponents; it is more convenient to use decimals. Then we have (using two significant figures in the exponent, for now)

$$10^{0.50} = 3.16.$$

Similarly, we find

$$10^{1/3} = 10^{.0.33} = \sqrt[3]{10} \approx 2.16.$$

These are numbers that can be worked out by standard, if somewhat obsolete methods; who these days ever learns how to take a cube root? (And if so, why?) However, suppose in some manner we work out the tenth root of ten; we would have

$$\sqrt[10]{10} = 10^{0.10} \approx 1.26.$$

With this we can calculate some other powers of ten, as follows:

$$10^{0.30} = (1.26)^3 \approx 2.0,$$
$$10^{0.60} = 10^{0.30} \times 10^{0.30} \approx 2.0 \times 2.0 = 4.0,$$
$$10^{0.70} = 10^{0.10} \times 10^{0.60} \approx 1.26 \times 4.0 = 5.0, \qquad (7)$$
$$10^{0.90} = 10^{0.60} \times 10^{0.30} \approx 2.0 \times 4.0 = 8.0.$$

Now we can define a new quantity, the *logarithm*, as follows: *the logarithm of a number is the power to which 10 must be raised to obtain that number.* For example, from Eqs. (3) and (4) above,

$$\begin{aligned}
\log 1000 &= 3, \\
\log 100 \ &= 2, \\
\log 10 \ \ &= 1, \\
\log 1 \ \ \ &= 0, \\
\log 0.1 \ \ &= -1,
\end{aligned} \qquad (8)$$

and so on. Also, from Eqs. (7), we have

$$\begin{aligned}
\log 2.0 &\approx 0.30, \\
\log 4.0 &\approx 0.60, \\
\log 5.0 &\approx 0.70, \\
\log 8.0 &\approx 0.90.
\end{aligned} \qquad (9)$$

In general, if we have

$$10^a = x, \qquad (10)$$

then

$$\log x = a. \qquad (11)$$

These logarithms have an extremely useful property. Suppose we have two arbitrary numbers x and y. They may be expressed as powers of ten, so we have

$$\begin{aligned}
x &= 10^a, \\
y &= 10^b.
\end{aligned} \qquad (12)$$

If we multiply these two equations together, we get

$$xy = 10^a \times 10^b = 10^{a+b}, \qquad (13)$$

by Eq. (5). Now from Eqs. (12), we have

$$\begin{aligned}
\log x &= a, \\
\log y &= b,
\end{aligned} \qquad (14)$$

and from Eq. (13) we have

$$\log xy = a+b. \qquad (15)$$

From Eqs. (10) and (11) we then get the following:

$$\log xy = \log x + \log y, \qquad (16)$$

that is, *the logarithm of the product of two numbers is equal to the sum of the logarithms of the numbers.* It follows that

$$\log x^n = n \log x. \qquad (17)$$

If we go through the same argument with the numbers x and y above but dividing instead of multiplying them and using Eq. (6) instead of Eq. (5) we get

$$\log (x/y) = \log x - \log y, \tag{18}$$

so that *the logarithm of the quotient of two numbers is the difference of the logarithms of the numbers,* the logarithm of the denominator being subtracted from that of the numerator.

It is the property of expressing multiplication and division in terms of addition and subtraction of logarithms that makes them so useful. Tables of logarithms of numbers have been calculated; with the help of such tables, complicated problems in multiplication and division can be reduced to problems of addition and subtraction. We do not need logarithms for this purpose, however, so we will not go into their use in calculations; we need them to find sound levels in decibels, and to calculate intervals in cents.

To find the logarithm of any number, we need only a table of logarithms of numbers from one to ten. Suppose we want the logarithm of 2000. We first express it in scientific notation, with one digit to the left of the decimal point; then by using Eq. (16), we have

$$\log 2000 = \log (2 \times 10^3) = \log 2 + \log (10)^3$$
$$= 0.30 + 3 = 3.30,$$

using the values of the logarithms given in Eqs. (8) and (9). Similarly,

$$\log 200 = \log (2 \times 10^2) = \log 2 + \log (10^2)$$
$$= 0.30 + 2 = 2.30,$$
$$\log 20 \ = 1.30,$$
$$\log 2 \ \ = 0.30,$$

and so on.

The logarithm then consists of two parts. The number to the right of the decimal point is the logarithm of the left-hand part of the product that expresses the number in scientific notation, this part having a value between one and ten. The number to the left of the decimal point in the logarithm is simply the exponent of the power of ten in the scientific notation product.

Numbers less than one have negative logarithms. For example, we have

$$\log (0.2) = \log (2 \times 10^{-1}) = \log 2 + \log 10^{-1}$$
$$= 0.30 - 1$$
$$= -0.70.$$

To find logarithms with sufficient accuracy for our purposes, Table I may be used. To find the logarithm of 3.5, for example, go down the left-hand column to 3., then across to the column directly under .5, and read the logarithm as 0.544.

TABLE I

Logarithms to three decimal places for numbers between 1 and 10

	.0	.1	.2	.3	.4	.5	.6	.7	.8	.9
1.	0.000	.041	.079	.114	.146	.176	.204	.230	.255	.279
2.	0.301	.322	.342	.362	.380	.398	.415	.431	.447	.462
3.	0.477	.491	.505	.519	.532	.544	.556	.568	.580	.591
4.	0.602	.613	.623	.633	.643	.653	.663	.672	.681	.690
5.	0.699	.708	.716	.724	.732	.740	.748	.756	.763	.771
6.	0.778	.785	.792	.799	.806	.813	.820	.826	.833	.839
7.	0.845	.851	.857	.863	.869	.875	.881	.886	.892	.898
8.	0.903	.908	.914	.919	.924	.929	.935	.940	.944	.949
9.	0.954	.959	.964	.968	.973	.978	.982	.987	.991	.996

Calculation of Cents

In Ch. 8 the frequency ratio ¢ for the interval of one cent was given as

$$¢ = 2^{1/1200}.$$

An interval of n cents is then given by the ratio

$$¢^n = 2^{n/1200},$$

since adding intervals is equivalent to multiplying their frequency ratios, as explained in Ch. 8. To find the number of cents n in any interval of frequency ratio R, we then have

$$2^{n/1200} = R. \qquad (19)$$

Now by taking the logarithm of each side, we have

$$\log\,(2^{n/1200}) = \log R,$$

and by using Eq. (17), we get

$$\frac{n}{1200} \log 2 = \log R.$$

The number of cents in the interval is then given by

$$n = 1200\,\frac{\log R}{\log 2}$$

$$= 3986 \log R. \qquad (20)$$

For example, let us find the number of cents in the just fifth. This has the frequency ratio $R = \frac{3}{2}$, so from Table I,

$$\log R = \log (\tfrac{3}{2}) = \log 3 - \log 2$$
$$= 0.477 - 0.301 = 0.176,$$

and consequently

$$n = 3986 \times 0.176$$
$$= 702 \text{ cents.}$$

As a further example of the usefulness of cents, suppose we want the sum of two intervals whose frequency ratios are R_1 and R_2. If n_1 and n_2 are the number of cents in these intervals, we have from Eq. (20) above

$$n_1 = 3986 \log R_1,$$
$$n_2 = 3986 \log R_2.$$

The frequency ratio R_s of the sum of the intervals is then the product $R_1 R_2$ of the individual frequency ratios. The number of cents n_s of the sum is then

$$n_s = 3986 \log R_s$$
$$= 3986 \log (R_1 R_2)$$
$$= 3986 [\log R_1 + \log R_2]$$
$$= n_1 + n_2.$$

Hence intervals expressed in cents may be added directly; this is a great convenience.

References

The list of references given below is furnished (a) to give the sources of certain statements made in the text, and (b) to give anyone interested in some particular topic a starting point for further reading. The list is not meant to be exhaustive.

JAES = *Journal of the Audio Engineering Society,* published by the Audio Engineering Society, 104 Liberty Street, Utica, N.Y. 13502.

JASA = *Journal of the Acoustical Society of America,* published by the American Institute of Physics, 335 E. 45th St., New York, N.Y. 10017.

JMT = *Journal of Music Theory,* published by the School of Music, Yale University, New Haven, Conn. 06520.

INTRODUCTION

[1] Georg von Bekesy, *Experiments in Hearing,* New York: McGraw-Hill, 1960, p. 8. A monumental work on the physiology and behavior of the ear.

CHAPTER ONE: THE FUNDAMENTAL PHYSICAL QUALITIES

[1] Hermann Helmholtz, *Sensations of Tone,* translated from the 4th German ed. of 1877 and material added by Alexander Ellis, New York: Dover Publications, 1954. This work is one of the landmarks in the literature of musical acoustics, as is obvious from the number of times it is cited in the text and in the references. On page 512, Helmholtz lists some lengths of the foot as used in various localities in Europe.

CHAPTER THREE: WAVES AND WAVE PROPAGATION

[1] Many measurements have been made of this number. For recent work, see H. C. Hardy, D. Telfair, and W. H. Pielemeier, *The Velocity of Sound in Air,* in JASA, XIII (1942), 226–33.
[2] L. J. Sivian, H. K. Dunn, and S. D. White, *Absolute Amplitudes and Spectra of Certain Musical Instruments and Orchestras,* in JASA, II (1931), 330–71.
[3] A. Bouhuys, *Sound Power Production in Wind Instruments,* in JASA, XXXVII (1965), 453–56.

CHAPTER FOUR: COMPLEX VIBRATIONS AND RESONANCE

[1] H. Levine and J. Schwinger, *On the Radiation of Sound from an Unflanged Circular Pipe,* in *Physical Review,* LXXIII (1948), 383–406.

CHAPTER FIVE: THE EAR

[1] Harvey Fletcher, *Speech and Hearing in Communication*, New York: D. Van Nostrand, 1953, p. 107.

[2] Bekesy, *op. cit.*, p. 100.

[3] *Ibid.*, p. 112.

[4] *Ibid.*, p. 462.

[5] L. J. Sivian and S. D. White, *On Minimum Audible Sound Fields*, in JASA, IV (1933), 288–321.

[6] M. Clark, Jr. and D. Luce, *Intensities of Orchestral Instrument Scales Played at Prescribed Dynamic Markings*, in JAES, XIII (1965), 151–57.

[7] Fletcher, *op. cit.*, p. 146.

[8] Recommendation R226 of the International Organization for Standardization, Geneva. These were standardized from curves given by D. W. Robinson and R. S. Dadson, *Equal-Loudness Relations for Pure Tones, and the Loudness Function*, in JASA, XXIX (1957), 1284–88. Earlier curves of the same kind, appearing in many books, were originally given by H. Fletcher and W. A. Munson, *Loudness, Its Definition, Measurement, and Calculation*, in JASA, V (1933), 82–108.

[9] S. S. Stevens, *Measurement of Loudness*, in JASA, XXVII (1955), 815–29; D. W. Robinson, *The Subjective Loudness Scale*, in Acustica, VII (1957), 217–33.

[10] J. C. Stevens and M. Guirao, *Individual Loudness Functions*, in JASA, XXXVI (1964), 2210–13.

[11] E. Zwicker, G. Flottorp, and S. S. Stevens, *Critical Bandwidth in Loudness Summation*, in JASA, XXIX (1957), 548–57.

[12] R. Plomp, *The Ear as a Frequency Analyzer*, in JASA, XXXVI (1964), 1628–36.

[13] Fletcher, *op. cit.*, p. 155.

[14] A. Pepinsky, *Masking Effects in Practical Instrumentation and Orchestration*, in JASA, XII (1941), 405–08.

[15] *Acoustic Trauma from Rock and Roll*, in High Fidelity, XVII/11 (Nov. 1967), 38–40.

[16] K. D. Kryter, W. D. Ward, J. D. Miller, and D. H. Eldredge, *Hazardous Exposure to Intermittent and Steady-State Noise*, in JASA, XXXIX (1966), 451–64.

CHAPTER SIX: TONE QUALITY

[1] U.S.A. Standards Institute (formerly American Standards Association), 10 East 40th St., New York, N.Y. 10016. See also the translator's footnote in Helmholtz, *op. cit.*, p. 25. His admonition against using the term *overtone* has gone largely unheeded for some eighty years. For further discussion of problems in terminology, see R. W. Young, *Modes, Nodes, and Antinodes*, in American Journal of Physics, XX (1952), 177–83.

[2] Helmholtz, *op. cit.*, p. 50 ff. For recent work demonstrating limitations on the ability to hear partials, see I. Pollack, *Ohm's Acoustical Law and Short-Term Auditory Memory*, in JASA, XXXVI (1964), 2340–45; also W. R. Thurlow and J. L. Rawlings, *Discrimination of a Number of Simultaneously Sounding Tones*, in JASA, XXXI (1959), 1332–36.

[3] Bekesy, *op. cit.*, p. 471.

[4] E. K. Chapin and F. A. Firestone, *Influence of Phase on Tone Quality and Loudness*, in JASA, V (1934), 173–80.

[5] R. C. Mathes and R. L. Miller, *Phase Effects in Monaural Perception*, in JASA, XIX (1947), 780–97.

[6] Under other conditions this will not necessarily be true; see J. H. Craig and L. A. Jeffress, *Effect of Phase on the Quality of a Two-Component Tone*, in JASA, XXXIV (1962), 1752–60.

[7] D. C. Miller, *The Science of Musical Sounds,* New York: Macmillan, 1926, p. 78 ff.

[8] C. S. McGinnis, H. Hawkins, and N. Sher, *Experimental Study of the Tone Quality of a Clarinet,* in JASA, XIV (1943), 228–37.

[9] M. Clark, Jr. and P. Milner, *Dependence of Timbre on Tonal Loudness Produced by Musical Instruments,* in JAES, XII (1964), 28–31.

[10] E. L. Saldanha and J. F. Corso, *Timbre Cues and the Identification of Musical Instruments,* in JASA, XXXVI (1964), 2021–26.

[11] M. D. Freedman, *Analysis of Musical Instrument Tones,* in JASA, XLI (1967), 793–806.

[12] F. A. Saunders, *Analyses of the Tones of a Few Wind Instruments,* in JASA, XVIII (1946), 395–407.

[13] P. Lehman, *Harmonic Structure of the Tone of the Bassoon,* in JASA, XXXVI (1964), 1649–53.

[14] R. H. Bolt, *Wanted—The Formant,* in JASA, XX (1948), 66.

[15] John Redfield, *Music: A Science and an Art,* New York: Alfred A. Knopf, 1926, p. 117.

[16] H. Fletcher, E. D. Blackham, and D. A. Christiansen, *Quality of Organ Tones,* in JASA, XXXV (1963), 314–25.

[17] James Jeans, *Science and Music,* London: Cambridge University Press, 1953, p. 240 ff. A similar discussion appears in many books, which imply that the missing fundamental is always heard. However, for exceptions see L. A. Jeffress, *The Pitch of Complex Tones,* in American Journal of Psychology, LIII (1940), 240–50.

[18] M. Lawrence and P. A. Yantis, *Onset and Growth of Aural Harmonics in the Overloaded Ear,* in JASA, XXVIII (1956), 852–58.

[19] Bekesy, *op cit.,* p. 335 f.

[20] J. L. Goldstein, *Auditory Nonlinearity,* in JASA, XLI (1966), 676–89.

[21] M. F. Meyer, *Aural Harmonics are Fictitious,* in JASA, XXIX (1957), 749.

[22] R. Plomp, *Beats of Mistuned Consonances,* in JASA, XLII (1967), 462–74.

[23] Helmholtz, *op. cit.,* p. 314.

[24] R. Plomp, *Detectability Threshold for Combination Tones,* in JASA, XXXVII (1965), 1110–23.

CHAPTER SEVEN: FREQUENCY AND PITCH

[1] W. B. Snow, *Audible Frequency Ranges of Music, Speech, and Noise,* in JASA, III (1931), 155–66.

[2] H. F. Olson, *Music, Physics, and Engineering,* New York: Dover Publications, 2nd ed., 1967, p. 250.

[3] S. S. Stevens, *The Relation of Pitch to Intensity,* in JASA, VI (1935), 150–54, gives curves relating pitch change to intensity. However, A. Cohen, *Further Investigation of the Effects of Intensity upon the Pitch of Pure Tones,* in JASA, XXXIII (1961), 1363–76, finds much less effect.

[4] H. Fletcher, *Loudness, Pitch, and the Timbre of Musical Tones and Their Relation to the Intensity . . . ,* in JASA, VI (1934), 59–69.

[5] W. B. Snow, *Change of Pitch with Loudness at Low Frequencies,* in JASA, VIII (1936), 14–19; D. Lewis and M. Cowan, *Influence of Intensity on the Pitch of Violin and Cello Tones,* in JASA, VIII (1936), 20–22.

[6] E. G. Shower and R. Biddulph, *Differential Pitch Sensitivity of the Ear,* in JASA, III (1932), 275–87.

[7] S. S. Stevens, J. Volkmann, and E. B. Newman, *A Scale for the Measurement of Psychological Magnitude Pitch,* in JASA, VIII (1937), 185–90.

[8] D. M. Neu, *A Critical Review of the Literature on Absolute Pitch,* in Psychological Bulletin, XLIV (1947), 249–66.

[9] C. H. Wedell, *A Study of Absolute Pitch,* in Psychological Bulletin, XXXVIII (1941), 547–48.

[10] A. Bachem, *Absolute Pitch,* in JASA, XXVII (1955), 1180–85.

[11] L. A. Jeffress, *Absolute Pitch,* in JASA, XXXIV (1962), 987.

[12] For an excellent review of the whole subject, see W. D. Ward, *Absolute Pitch,* in *Sound* II/3 (May-June 1963), 14–21; II/4 (July-Aug. 1963), 33–41.

CHAPTER EIGHT: INTERVALS, SCALES, TUNING, AND TEMPERAMENT

[1] Jeans, *op. cit.,* p. 164.

[2] J. M. Barbour, *Tuning and Temperament,* East Lansing: Michigan State College Press, 2nd ed., 1953, p. 25 ff.

[3] *Ibid.,* p. 107 ff.

[4] Acoustical Society of America, *Report of Standards Committee, Sec. 4014,* in JASA, II (1931), 311–24. In fairness, it should be said that this definition was subsequently changed: see JASA, IX (1937), 60–71.

[5] Helmholtz, *op. cit.,* p. 316 ff.

[6] Records are available for listening in meantone and just tunings: J. M. Barbour and F. Kuttner, *Meantone Temperament in Theory and Practice,* Theory Series A–2; *Theory and Practice of Just Intonation,* Theory Series A–3; published by Musurgia Records, 309 West 104th St., New York 10025, N.Y.

[7] Helmholtz, *op. cit.,* p. 466 ff. For a more recent effort, see C. Williamson, *Keyboard Instrument in Just Intonation,* in JASA, XV (1944), 173–75.

[8] Helmholtz, *op. cit.,* p. 429 f.

[9] Helmholtz, *op. cit.,* p. 325.

[10] P. C. Greene, *Violin Intonation,* in JASA, IX (1937), 43–44; J. F. Nickerson, *Intonation of Solo and Ensemble Performance of the Same Melody,* in JASA, XXI (1949), 593–95.

[11] W. Lottermoser and J. Meyer, *Frequenzmessungen an Gesungen Akkorden,* in *Acustica,* X (1960), 181–84.

[12] D. W. Martin, *Musical Scales Since Pythagoras,* in *Sound* I/3 (May–June 1962), 22–24; W. D. Ward and D. W. Martin, *Psychophysical Comparison of Just Tuning and Equal Temperament in Sequences of Individual Tones,* in JASA, XXXIII (1961), 586–88.

[13] A. L. Leigh Silver, *Equal Beating Chromatic Scale,* in JASA, XXIX (1957), 476–81; T. E. Simonton, *New Integral Ratio Chromatic Scale,* in JASA, XXV (1953), 1167–75.

[14] For a recent theory, see P. Boomsliter and W. Creel, *Extended Reference—an Unrecognized Dynamic in Melody,* in JMT, VII (1963), 2–22.

[15] Helmholtz, *op. cit.,* p. 495 ff.

[16] R. W. Young, *Why an International Tuning Frequency,* in JASA, XXVII (1955), 379–80.

[17] G. Hendricks, *The Case of the Disappearing High C's,* in *Etude,* LXVII/2 (1949), 77.

[18] F. A. Saunders, *A Scientific Search for the Secret of Stradivarius,* in *Journal of the Franklin Institute,* CCIX (Jan. 1940), 1–20.

[19] R. W. Young, *Terminology for Logarithmic Frequency Units,* in JASA, XI (1939), 134–39.

[20] R. W. Young, *Dependence of Tuning of Wind Instruments on Temperature,* in JASA, XVII (1946), 187–91.

[21] O. L. Railsback, *A Chromatic Stroboscope,* in JASA, IX (1937), 37–42. This instrument is sold by C. G. Conn, Ltd., Elkhart, Indiana, 46515, under the trade name "Stroboconn."

[22] O. J. Murphy, *Measurements of Orchestral Pitch,* in JASA, XII (1941), 395–98.

[23] Helmholtz, *op. cit.,* p. 182 ff.

[24] H. Moran and C. C. Pratt, *Variability of Judgments on Musical Intervals,* in *Journal of Experimental Psychology,* IX (1926), 492–500; C. R. Shackford, *Some*

Aspects of Perception, in JMT, V (1961), 295–303; VI (1962), 66–90 and 295–303. See particularly the last of these, p. 300, for a chart showing ranges of intervals.

25 R. Plomp and W. J. M. Levelt, *Tonal Consonance and Critical Bandwidth,* in JASA, XXXVIII (1965), 548–60.

26 For a recent theory, see P. Boomsliter and W. Creel, *The Long Pattern Hypothesis in Harmony and Hearing,* in JMT, V (1961), 2–31.

27 J. F. Corso, *Absolute Judgments of Musical Tonality,* in JASA, XXIX (1957), 138–44.

28 Helmholtz, *op cit.,* p. 551.

CHAPTER NINE: AUDITORIUM AND ROOM ACOUSTICS

1 I. Rudnick, *Propagation of an Acoustic Wave Along a Boundary,* in JASA, XIX (1947), 348–56.

2 Wallace Clement Sabine, *Collected Papers on Acoustics,* New York: Dover Publications, 1964.

3 Philip M. Morse, *Vibration and Sound,* New York: McGraw-Hill, 1948, p. 386.

4 Additional curves may be found in Leslie L. Doelle, *Acoustics in Architectural Design,* Ottawa, Canada: National Research Council, Division of Building Research, 1965, p. 136.

5 *Ibid.,* p. 101 ff.

6 D. Olynyk and T. D. Northwood, *Comparison of Reverberation-Room and Impedance-Tube Absorption Measurements,* in JASA, XXXVI (1964), 2171–74.

7 R. N. Lane and J. Botsford, *Total Sound Absorption for Upholstered Theater Chairs with Audience,* in JASA, XXIV (1952), 125–26; also Doelle, *op. cit.,* p. 106.

8 R. W. Young, *Sound Absorption in Air in Rooms,* in JASA, XXIX (1957), 311.

9 A. H. Benade, *Resonance Absorption Cross-Section of a Pipe Organ,* in JASA, XXXVIII (1965), 780–89.

10 L. Beranek, *Acoustic Measurements,* New York: John Wiley and Sons, 1956, p. 862.

11 L. Beranek, *Music, Acoustics, and Architecture,* New York: John Wiley and Sons, 1962, p. 543.

12 L. Beranek, *Audience and Seat Absorption in Large Halls,* in JASA, XXXII (1960), 661–70.

13 R. N. Lane, *Absorption Characteristics of Upholstered Theater Chairs and Carpet as Measured in Two Auditoriums,* in JASA, XXVIII (1956), 101–05.

14 An improvement on this method is described by M. R. Schroeder, *New Method of Measuring Reverberation Time,* in JASA, XXXVII (1965), 409–12.

15 This device is based on the principle outlined in the preceding reference. It has been demonstrated by the Bell Telephone Laboratories; a published description is not yet available.

16 T. J. Schultz, *Using Music to Measure Reverberation Time,* in *Gravesano Review,* 27/28 (1966), 120–22.

17 D. Pinkham, *Catacoustical Measures,* New York: C. F. Peters, 1962.

18 P. Veneklasen and J. P. Christoff, *The Seattle Opera House—Acoustical Design,* in JASA, XXXVI (1964), 903–10.

19 V. O. Knudsen, *Review of Architectural Acoustics During the Past Twenty-Five Years,* in JASA, XXVI (1954), 646–50.

20 T. J. Schultz and B. G. Watters, *Propagation of Sound Across Audience Seating,* in JASA, XXXVI (1964), 885–96.

21 Jeans, *op. cit.,* p. 211.

22 Beranek, *Music, Acoustics, and Architecture,* p. 63.

23 L. Beranek, F. R. Johnson, T. J. Schultz, and B. G. Watters, *Acoustics of Philharmonic Hall, New York, During its First Session,* in JASA, XXXVI (1964), 1247–62.

[24] M. R. Schroeder, B. S. Atal, G. M. Sessler, and J. E. West, *Acoustical Measurements in Philharmonic Hall* (*N.Y.*), in JASA, XL (1966), 434–40.

[25] R. W. Leonard, L. P. Delsasso, and V. O. Knudsen, *Diffraction of Sound by an Array of Rectangular Reflective Panels*, in JASA, XXXVI (1964), 2328–33.

[26] T. J. Schultz, *Acoustics of the Concert Hall*, in *Institute of Electrical and Electronics Engineers Spectrum*, II (June 1965), 56–57.

[27] R. S. Shankland, *Acoustics of N.Y. Philharmonic Hall*, in JASA, XXXV (1963), 725–26; R. S. Lanier, *What Happened at Philharmonic Auditorium?* in *Architectural Forum*, CXIX (Dec. 6, 1963), 118–23.

[28] E. Krauth and R. Büchlein, *Model Tests in Architectural Acoustics*, in *Gravesano Review*, 27/28 (1966), 155–60.

[29] R. N. Lane and E. E. Mikeska, *Study of Acoustical Requirements for Teaching Rooms and Practice Rooms in Music School Buildings*, in JASA, XXVII (1955), 1087–91; L. S. Goodfriend, *Acoustics for School Music Departments*, in *Sound*, II/1 (Jan.–Feb. 1963), 28–32.

[30] W. R. MacLean, *On the Acoustics of Cocktail Parties*, in JASA, XXXI (1959), 79–80.

[31] L. Beranek, *Sound Systems for Large Auditoriums*, in JASA, XXVI (1954), 661–75.

[32] H. Haas, *Über den Einfluss eines Einfachechos auf die Hörsamkeit von Sprache*, in *Acustica*, 1 (1951), 49–58.

[33] C. P. Boner and C. R. Boner, *Minimising Feedback in Sound Systems and Room-Ring Modes with Passive Networks*, in JASA, XXXVII (1965), 131–35.

[34] H. Burris-Meyer, *Control of Acoustic Conditions on the Concert Stage*, in JASA, XII (1941), 335–37; D. W. Martin, *Supplementary Sound for Opera*, in *Sound* I/1 (Jan.–Feb. 1962), 25–33.

CHAPTER TEN: THE STRING INSTRUMENTS

[1] For more discussion of violin string motion, see J. C. Schelleng, *The Bowed String*, in *American String Teacher*, XVII (Summer 1967), 15–19.

[2] F. A. Saunders, *The Mechanical Action of Violins*, in JASA, IX (1937), 81–98; *Violins Old and New—An Experimental Study*, in *Sound* I/4 (July–Aug. 1962), 7–15.

[3] See above, Ch. 8, ref. 18.

[4] C. M. Hutchins, *The Physics of Violins*, in *Scientific American*, CCVII (Nov. 1962), 78–93.

[5] F. A. Saunders, *Recent Work on Violins*, in JASA, XXV (1953), 491–98.

[6] C. M. Hutchins, *Founding a Family of Fiddles*, in *Physics Today*, XX/2 (Feb. 1967), 23–37.

[7] H. Backhaus and G. Weymann, *New Results in Research on Violins*, in JASA, XI (1940), 490–92.

[8] C. M. Hutchins, A. S. Hopping, and F. A. Saunders, *Subharmonics and Plate Tap Tones in Violin Acoustics*, in JASA, XXXII (1960), 1443–49; R. B. Watson, W. J. Cunningham, and F. A. Saunders, *Improved Techniques in the Study of Violins*, in JASA, XII (1941), 399–402.

[9] See preceding references.

[10] See ref. 2 above.

[11] H. Meinel, *Regarding the Sound Quality of Violins and a Scientific Basis for Violin Construction*, in JASA, XXIX (1957), 817–22.

[12] J. C. Schelleng, *On the Physical Effects of Violin Varnish*, in *Catgut Acoustical Society Newsletters*, 6 (Nov. 1966), 7 (May 1967), 8 (Nov. 1967).

[13] R. B. Abbott and G. H. Purcell, *Physical Properties of Wood for Violin Construction*, in JASA, XIII (1941), 54–55.

[14] See ref. 5 above.

[15] E. D. Blackham, O. H. Geersten, and H. Fletcher, *Quality of Violin, Viola, Cello, and Bass-Viol Tones,* in JASA, XXXVII (1965), 851–63.

[16] F. A. Saunders, *The Mechanical Action of Instruments of the Violin Family,* in JASA, XVII (1946), 169–86.

[17] See preceding reference.

[18] See ref. 6 above.

[19] J. C. Schelleng, *The Violin as a Circuit,* in JASA, XXXV (1963), 326–38.

[20] See refs. 4 and 6 above; also C. M. Hutchins and J. C. Schelleng, *A New Concert Violin,* in JAES, XV (1967), 432–36.

[21] J. C. Schelleng, *Adjusting the Wolftone Suppressor,* in *American String Teacher,* XVII (Winter 1967), 9.

[22] Address for information, contributions, etc.: 112 Essex Ave., Montclair, N.J. 07042.

CHAPTER ELEVEN: THE WOODWIND INSTRUMENTS, AND OTHERS

[1] A. Powell, *On the Edgetone,* in JASA, XXXIII (1961), 395–409; G. B. Brown, *Edge Tones,* in *Proceedings of the Physical Society of London,* XLIX (1937), 493–507.

[2] Anthony Baines, *Woodwind Instruments and Their History,* New York: W. W. Norton, 1957, p. 171 f.

[3] *Ibid.,* p. 52 ff.; also Theobald Boehm, *The Flute and Flute Playing,* trans. by D. C. Miller, New York: Dover Publications, 1964.

[4] Baines, *op. cit.,* p. 123 ff.; also F. Geoffrey Rendall, *The Clarinet,* London: Ernest Benn, 2nd ed., 1957.

[5] J. Backus, *Vibrations of the Reed and the Air Column in the Clarinet,* in JASA, XXXIII (1961), 806–09.

[6] Philip Bate, *The Oboe,* London: Ernest Benn, 1956.

[7] Lyndesay G. Langwill, *The Bassoon and the Contrabassoon,* New York: W. W. Norton, 1965.

[8] Baines, *op. cit.,* pp. 143–47.

[9] See above, Ch. 3, ref. 3.

[10] A. H. Benade, *On Woodwind Instrument Bores,* in JASA, XXXI (1959), 137–46.

[11] J. W. Coltman, *Resonance and Sounding Frequencies in the Flute,* in JASA, XL (1966), 99–107.

[12] J. Backus, *Small Vibration Theory of the Clarinet,* in JASA, XXXV (1963), 305–13.

[13] See ref. 11 above.

[14] R. W. Young and J. C. Webster, *Tuning of Musical Instruments: The Clarinet,* in *Gravesano Review,* IV (1958), 182–86. For a discussion of intonation in general, see Donald Stauffer, *Intonation Deficiencies of Wind Instruments,* Washington, D.C.: Catholic University of America Press, 1954.

[15] D. C. Miller, *The Influence of the Material of Wind Instruments on Tone Quality,* in *Science,* XXIX (1909), 161–71.

[16] S. E. Parker, *Analyses of the Tones of Wooden and Metal Clarinets,* in JASA, XIX (1947), 415–19.

[17] Rendall, *op. cit.,* p. 11 ff.; Bate, *op. cit.,* p. 129 ff.

[18] J. Backus, *Effect of Wall Material on the Steady State Tone Quality of Woodwind Instruments,* in JASA, XXXVI (1964), 1881–87.

[19] D. M. A. Mercer, *Voicing of Flue Organ Pipes,* in JASA, XXIII (1951), 45–54; *Effect of Voicing Adjustments on the Tone Quality of Organ Pipes,* in *Acustica,* IV (1954), 237–39.

[20] W. Lottermoser, *Der Einfluss des Materials von Orgel-Metallpfeifen auf ihre Tongebung,* in *Akustische Zeitschrift,* II (1937), 129–34; also ref. 15 above.

[21] William H. Barnes, *The Contemporary American Organ*, Glen Rock, N.J.: J. Fischer & Bro., 8th ed., 1964.

[22] E. von Glatter-Götz, *Der Einfluss des Wandmaterials von Orgelpfeifen auf Klangfarbe und Lautstärke*, in *Zeitschrift für Instrumentenbau*, LV (1935), 96–99, finds no effect; ref. 20 above disagrees.

[23] C. P. Boner and R. B. Newman, *Effect of Wall Material on the Steady State Acoustic Spectra of Flue Pipes*, in JASA, XII (1940), 83–89.

[24] J. Backus and T. C. Hundley, *Wall Vibrations in Flue Organ Pipes and Their Effect on Tone*, in JASA, XXXIX (1966), 936–45.

[25] George A. Audsley, *The Art of Organ Building*, New York: Dover Publications, 1965; also ref. 21 above.

[26] Fletcher, *Speech and Hearing*, p. 53.

[27] W. T. Bartholomew, *The Paradox of Voice Teaching*, in JASA, XI (1939), 446–50.

[28] S. K. Wolf, D. Stanley, and W. J. Sette, *Quantitative Studies on the Singing Voice*, in JASA, VI (1935), 255–66.

[29] G. A. Sacerdote, *Researches on the Singing Voice*, in *Acustica*, VII (1957), 61–68. For earlier work on this and many other musical problems, see Carl E. Seashore, *The Psychology of Music*, New York: McGraw-Hill, 1938.

[30] W. T. Bartholomew, *A Physical Definition of "Good Voice Quality" in the Male Voice*, in JASA, VI (1934), 25–33; A. Bjørklund, *Analysis of Soprano Voices*, in JASA, XXXIII (1961), 575–82.

[31] D. Lewis, *Vocal Resonance*, in JASA, VIII (1937), 91–99.

CHAPTER TWELVE: THE BRASS INSTRUMENTS

[1] Philip Bate, *The Trumpet and Trombone*, New York: W. W. Norton, 1966, p. 85.

[2] H. W. Henderson, *Experimental Study of Trumpet Embouchure*, in JASA, XIV (1942), 58–64.

[3] J. C. Webster, *Electrical Method of Measuring Intonation of Cup-Mouthpiece Instruments*, in JASA, XIX (1947), 902–06.

[4] W. T. Cardwell, *Working Theory of Trumpet Air-Column Design*, in JASA, XL (1966), 1252.

[5] Arthur H. Benade, *Horns, Strings, and Harmony*, Garden City, N.Y.: Doubleday Anchor Books, 1960, p. 177.

[6] J. C. Webster, *Internal Tuning Differences Due to Players and Taper of Trumpet Bells*, in JASA, XXI (1949), 208–14.

[7] Bate, *The Trumpet and Trombone*, p. 167.

[8] J. Redfield, *Minimising Discrepancies of Intonation in Valve Instruments*, in JASA, III (1931), 292–96; R. W. Young, *Optimum Lengths of Valve Tubes for Brass Wind Instruments*, in JASA, XLII (1967), 224–35.

[9] For a grotesque example, see Bate, *The Trumpet and Trombone*, p. 175.

[10] *Grove's Dictionary of Music and Musicians*, London: Macmillan, 5th ed., 1954.

[11] Gunther Schuller, *Horn Technique*, New York: Oxford University Press, 1967.

[12] R. Morley-Pegge, *The French Horn*, London: Ernest Benn, 1960.

[13] *Ibid.*, p. 137; the same error is stated in other books written by musicians, such as Walter Piston, *Orchestration*, New York: W. W. Norton, 1955, p. 235. Stauffer (Ch. 11, ref. 14) gives the correct explanation, p. 60.

[14] See ref. 10 above.

[15] See above, Ch. 3, ref. 3.

[16] D. W. Martin, *Directivity and Acoustic Spectra of Brass Wind Instruments*, in JASA, XIII (1942), 309–13.

[17] J. E. Ancell, *Sound Pressure Spectra of a Muted Cornet*, in JASA, XXXII (1960), 1101–04.

[18] D. W. Martin, *Lip Vibrations in a Cornet Mouthpiece*, in JASA, XIII (1942) 305–08.

[19] T. H. Long, *Performance of Cup-Mouthpiece Instruments*, in JASA, XIX (1947), 892–901.

[20] H. D. Knauss and W. J. Yeager, *Vibrations of the Walls of a Cornet*, in JASA, XIII (1941), 160–62.

CHAPTER THIRTEEN: THE PIANO

[1] E. D. Blackham, *Physics of the Piano*, in Scientific American, CCXIII/6 (Dec. 1965), 88–99.

[2] P. H. Bilhuber and C. A. Johnson, *Influence of the Soundboard on Piano Tone Quality*, in JASA, XI (1940), 311–20.

[3] William Braid White, *Piano Tuning and Allied Arts*, Boston: Tuner's Supply Co., 5th ed., 1946, p. 190 ff.

[4] Helmholtz, *op. cit.*, p. 78 ff. One of the latest repetitions may be found in Olson (Ch. 7, ref. 2), p. 126.

[5] O. H. Schuck and R. W. Young, *Observations on the Vibrations of Piano Strings*, in JASA, XV (1943), 1–11.

[6] D. W. Martin, *Decay Rates of Piano Tones*, in JASA, XIX (1947), 535–41.

[7] R. E. Kirk, *Tuning Preferences for Piano Unison Groups*, in JASA, XXXI (1959), 1644–48.

[8] See ref. 1 above.

[9] R. W. Young, *Inharmonicity of Plain Wire Piano Strings*, in JASA, XXIV (1952), 267–73; H. Fletcher, *Normal Vibration Frequencies of a Stiff Piano String*, in JASA, XXXVI (1964), 203–09.

[10] See ref. 5 above.

[11] H. Fletcher, E. D. Blackham, and R. Stratton, *Quality of Piano Tones*, in JASA, XXXIV (1962), 749–61.

[12] F. Miller, Jr., *Proposed Loading of Piano Strings for Improved Tone*, in JASA, XXI (1949), 318–22.

[13] See ref. 3 above.

[14] W. B. White, *Practical Tests for Determining the Accuracy of Piano Tuning*, in JASA, IX (1937), 47–50.

[15] R. W. Young, *Influence of Humidity on the Tuning of a Piano*, in JASA, XXI (1949), 580–85.

[16] See ref. 5 above.

[17] See ref. 5 above.

[18] D. W. Martin and W. D. Ward, *Subjective Evaluation of Musical Scale Temperament in Pianos*, in JASA, XXXIII (1961), 582–88.

[19] Tobias Matthay, *The Act of Touch in All Its Diversity*, London: Longmans, Green, 1911, p. 50. The nonsense content of this book is very high. On p. 262 are described 42 distinct piano touches! The same comment applies to Matthay's other books.

[20] W. B. White, *The Human Element in Piano Tone Production*, in JASA, I (1930), 357–65.

[21] H. C. Hart, M. W. Fuller, and W. S. Lusby, *A Precision Study of Piano Touch and Tone*, in JASA, VI (1934), 80–94.

[22] Otto Ortmann, *The Physical Basis of Piano Touch and Tone*, New York: Dutton, 1925; Dutton Paperback D104, 1962.

CHAPTER FOURTEEN: THE PERCUSSION INSTRUMENTS

[1] H. C. Hardy and J. E. Ancell, *Comparison of the Acoustical Performance of Calfskin and Plastic Drumheads*, in JASA, XXXIII (1961), 1391–95.

[2] Henry W. Taylor, *The Timpani*, London: John Baker, 1964.

[3] J. Obata and T. Tesima, *Experimental Studies on the Sound and Vibration of Drums*, in JASA, VI (1935), 267–74.

CHAPTER FIFTEEN: THE ELECTRONIC PRODUCTION OF SOUND

[1] A. M. Springer, *A Pitch Regulator and Information Changer*, in Gravesano Review, 11/12 (1958), 7–9.

[2] For practical discussions of high-fidelity systems, see Edgar Villchur, *Reproduction of Sound*, New York: Dover Publications, 1965; Steven Hahn, *Hi-Fi Handbook*, New York: Thomas Y. Crowell, 1962.

[3] For a technical discussion of loudspeaker problems, see N. W. McLachlan, *Loudspeakers*, New York: Dover Publications, 1960.

[4] R. E. Kirk, *Learning, A Major Factor Influencing Preferences in High Fidelity Reproducing Systems*, in JASA, XXVIII (1956), 1113–16.

[5] J. K. Hilliard, *History of Stereophonic Sound Reproduction*, in *Proceedings of the Institute of Radio Engineers*, L (1962), 776–80.

[6] H. Bode, *Sound Synthesizer Creates New Musical Effects*, in Electronics, XXXIV (Dec. 1, 1961), 33–37.

[7] D. W. Martin and A. F. Knoblaugh, *Loudspeaker Accessory for the Production of Reverberant Sound*, in JASA, XXVI (1954), 676–78.

[8] A. Douglas, *Some Electronic Extensions to Music Generating Systems*, in Electronic Engineering, XXXV (1963), 726–31; H. Hearne, *Electronic Production of Percussive Sounds*, in JAES, IX (1961), 270–71.

[9] Richard H. Dorf, *Electronic Musical Instruments*, Mineola, N.Y.: Radio Magazines, Inc., 1954; Alan Douglas, *The Electronic Musical Instrument Manual*, New York: Pitman, 3rd ed., 1957.

[10] H. F. Olson, H. Belar, and J. Timmens, *Electronic Music Synthesis*, in JASA, XXXII (1960), 311–19. A later version is described in M. Babbitt, *An Introduction to the RCA Synthesizer*, in JMT, VIII (1964), 251–65. For a recording of synthesizer music, hear RCA Victor LM 1922 (now deleted); also Columbia MS–6566 and 7051.

[11] V. Ussachevsky, *The Process of Experimental Music*, in JAES, VI (1958), 202–07.

[12] Some installations are described by R. M. Voss, *Brandeis University Experimental Music Studio*, in JAES, XIII (1965), 65–68, and L. A. Hiller, Jr., *An Integrated Electronic Music Console*, in JAES, XIII (1965), 142–50.

[13] For a listing of available records, see Hugh Davies, *International Electronic Music Catalog*, in *Electronic Music Review*, Nos. 2/3 (April/July 1967), pp. 242–75.

[14] *Die Reihe*, Universal Edition, Vienna; English translation by Theodore Presser Co., Bryn Mawr, Pa. This journal has ceased publication. A new journal has recently appeared: *Electronic Music Review*, published by The Independent Electronic Music Center, Trumansburg, N.Y. 14886.

[15] J. Backus, *Die Reihe: A Scientific Evaluation*, in Perspectives of New Music, I/1 (1962), 160–71.

[16] M. V. Mathews, *The Digital Computer as a Musical Instrument*, in Science, CXLII (1963), 553–57.

[17] J. R. Pierce, M. V. Mathews, and J. C. Risset, *Further Experiments on the Use of a Computer in Connection with Music*, in Gravesano Review, 27/28 (1966), 92–97.

[18] J. C. Tenney, *Sound Generation by Means of a Digital Computer*, in JMT, VII (1963), 24–70.

[19] J. R. Pierce, *The Computer as a Musical Instrument*, in JAES, VIII (1960), 139–40.

[20] *Music from Mathematics*, Bell Telephone Laboratories Record #122227.

[21] J. L. Divilbis, *Real-Time Generation of Music with a Digital Computer,* in JMT, VIII (1964), 99–111.

[22] L. A. Hiller, Jr. and L. M. Isaacson, *Musical Composition with a High Speed Digital Computer,* in JAES, VI (1958), 154–60.

[23] *Iliac Suite for String Quartet* (1957) and *Computer Cantata* (1963), Heliodor S 25053.

[24] H. F. Olson and H. Belar, *Aid to Music Employing a Random Probability System,* in JASA, XXXIII (1961), 1163–70. For an illuminating commentary, see J. M. Barbour, *Comment on "Aid to Music . . . ,"* in JASA, XXXIV (1962), 128–29.

[25] L. A. Hiller Jr., *Musical Applications of Electronic Digital Computers,* in *Gravesano Review,* 27/28 (1966), 62–72.

[26] A general discussion of work needed is given by R. W. Young, *Some Problems for Postwar Musical Acoustics,* in JASA, XVI (1945), 103–07. (The problems of twenty years ago are still problems today.)

[27] R. W. Young, *Twenty-Five Years of Musical Acoustics,* in JASA, XXVI (1954), 955–59.

[28] H. Fletcher, *An Institute for Musical Sciences—A Suggestion,* in JASA, XIX (1947), 527–31.

Index